Princess Pale Moon
American Indian Heritage Foundation
6051 Arlington Blvd.
Falls Church, Va. 22044

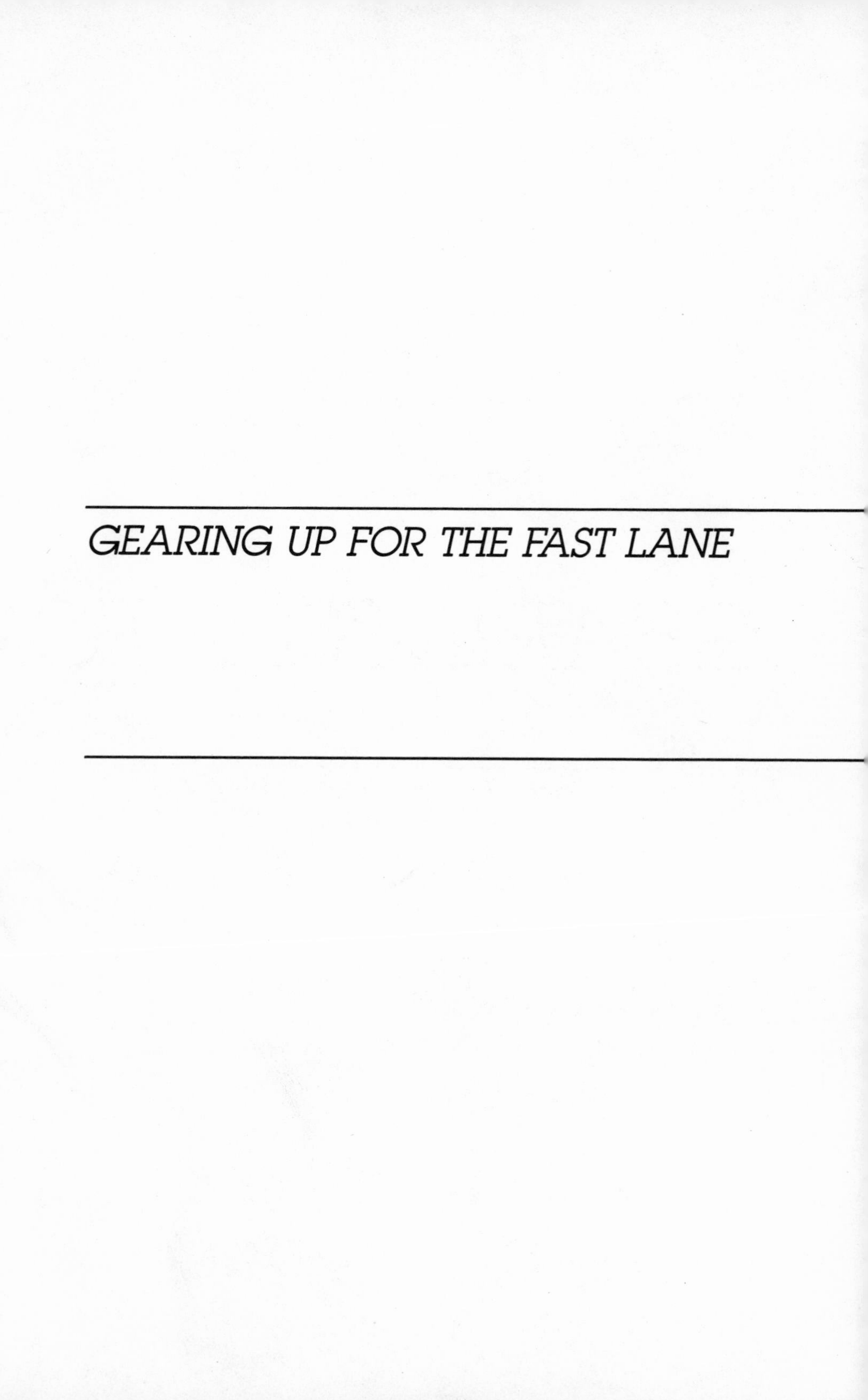

GEARING UP FOR THE FAST LANE

GEARING UP FOR

THE FAST LANE

New Tools for Management in a High-Tech World

Deborah S. Bright

RANDOM HOUSE BUSINESS DIVISION
NEW YORK

First Edition
987654321

All inquiries should be addressed to Random House, Inc., 201 East 50th Street, New York, N.Y. 10022. Published in the United States by Random House, Inc., and simultaneously in Canada by Random House of Canada Limited, Toronto.

Library of Congress Cataloging-in-Publication Data

Bright, Deborah.
Gearing up for the fast lane.

Includes index.
1. High technology industries—Management. I. Title.
HD62.37.B75 1985 621.381'7'068 85-8375
ISBN 0-394-55003-X

Manufactured in the United States of America

CONTENTS

FOREWORD

Just over twenty years ago, I had my first encounter with emerging technology as an engineering graduate student working for NASA. Even in that future-oriented environment, the current concept of high technology as a social force was still years away, particularly as it might affect the practice of management. Even managers in technology-based companies routinely used no technology higher than a pencil and paper. In many industries, business had been transacted and employees managed in the same ways for generations.

Twenty years later, revolutions in transportation and communications have made rapid change and intense worldwide competition the rule even for traditional industries. Managers in these companies increasingly find traditional management styles inadequate to the massive challenges they are facing. The pressure for innovation in business and in management methods has become inescapable. The fact that technology alters the conduct of business is not new or unexpected. What was *not* expected, and what Dr. Bright is revealing in *GEARING UP FOR THE FAST LANE*, are the significant new management concepts that the high-tech world developed to cope with its own breakneck pace.

The industry that I know best—personal computers—represents probably the most visible and explosive of the new breed of high-tech industries. It didn't even exist in the mid-1970s. Ten years later, by the mid-1980s, it was estimated to be an $11 billion business, with even more growth anticipated. The power of personal computer technology grew just as rapidly. Business personal computers on the market today have capacities hundreds of times greater than those of the first microcomputers introduced ten years ago.

The growth of the personal computer industry is impressive, but the turnover among its participants is just as impressive. At times it has resembled an old-fashioned gold rush, with hundreds of companies having come and gone in just a few years. At any given time, the industry had more than 150 competing manufacturers, and the software companies were almost uncountable. Yet in a very short time, the principal participants have dwindled to a handful, and their number is still shrinking.

Was success or failure in the PC boomtown based on technology alone? Probably not. Technology has changed rapidly, but within manageable limits. The realities of business required that personal computers be similar enough to be compatible. If technology wasn't the single determining factor, then what has made the difference among these high-tech companies? More importantly, what can we learn from those that made it, and those that didn't?

There is really only one company I know well enough to use as an example—COMPAQ® Computer Corporation. In many ways, COMPAQ® has been a laboratory for the principles expressed by Dr. Bright. It was born at the beginning of the most competitive period in the industry. The company was built from the ground up, by people from many high-tech companies, in a business market that was growing and changing more rapidly than any in history. In building the company, we deliberately chose the same type of people Dr. Bright interviewed—experienced, successful high-tech managers from around the world. They had to act quickly, drawing on their experience to apply the best managerial techniques from their varied high-tech backgrounds. The result? A company that grew even faster than the personal computer industry itself, and in less than three years became a strong, viable competitor with some of the giants in electronics and computers.

The managers Dr. Bright interviewed have shared an experience similar to ours at COMPAQ®. Whether from emerging or established companies, they have seen their own creativity interacting with that of their co-workers in unexpected ways. They have created both new products and new ways of working together. Most important, they have learned that success in the volatile, intensely competitive high-tech world depends on the creativity of every employee, not just that of a single employee or of a small group.

Genuinely good new ideas seem obvious once we become familiar with them, such as Buckminster

Fuller's idea of synergy. *GEARING UP FOR THE FAST LANE* presents us with ideas of that sort. From her research among scores of successful high-tech managers, Dr. Bright has refined concepts that will seem obvious after using them. The greatest contribution her book makes is in presenting management practices developed empirically and learned informally in a structured and well-organized format. She makes them available, not only to high-tech managers, but also to managers in traditional firms. Dr. Bright gives managers practical guidelines for adopting these ideas in their own organizations—and for encouraging the same enthusiasm and commitment that characterizes success in the high-tech world.

GEARING UP FOR THE FAST LANE directly addresses basic issues faced by most U.S. companies today. As the book points out, managers in most industries are moving into the fast lane whether they want to or not—they have no choice. Managers are having to confront issues ranging from the increasing use of high technology in their businesses to the intensifying pressures of global competition.

It is not only possible to survive, but to prosper, in this new environment. The emerging "high-tech economy" offers exciting opportunities, but it requires managers to do nothing less than elicit exceptional performance from themselves and others. The many successful managers represented in *GEARING UP FOR THE FAST LANE* achieve exceptional performance. Dr. Bright not only tells us how to do it, but gives us a blueprint for encouraging the same performance from the millions of new high-tech workers that are needed to meet these challenges.

Speaking from my own experience, there is

another very good reason for applying the management principles Dr. Bright has described. They can lead to some of the most rewarding experiences of our lives, bringing the elusive qualities of genuine satisfaction, and even fun, to our jobs.

ROD CANION
President and Chief Executive Officer
COMPAQ® Computer Corporation

ACKNOWLEDGEMENTS

I would like to thank the following people for their participation.

Jerry Ade
Bill Allen
Hector Aponte
Tom Ashley
Roger Attanas
Steve Bailey
Richard Baker
Guy Barnicoat
Alan Bass
Barry Beder
Peter Beham
Larry Biggs
Myron W. Black
Zeb Black
Timothy Blodgett
Marlene Blum
Paul Bourdreau
Miles Boyer
Mike Brady
Barbara Brams
Marge Brockman
Allen Brunner
Richard Burke
Ronald Burke
Dr. William Burlant
Mike Burnes
Brian Burns
Kay Burns
Bill Burton
Bill Byham
Joyce Campbell
Jim Carbonara

Don Carlton
Mary Carson
Paula Carter
Bev Cella
Mary Cernik
Henry Chamberlain
Ken Chiara
Gary Christel
Pete Chylko
John Cindric, Jr.
Pat Clough
Bill Cobb
Logan Cockrum
Jim Coleman
Wendy Coles
Lance Cooper
Jerry Cordrey
Juliana Cornish
John Cosgrove
Dennis Craig
William Creighton
Barry Crick
James Croll
Tom Cummings
Walter Curran
Rick Custer
Ted Davenport
Bob Davis
Mel Davis
Chester Delaney
William Densham
Benjamin Dent
Carl D'Oyley
Gerald DeVore
Dan DeZwarte
F. J. Dietrich
Judy Divita
William Draper
Tom Dunne
Lew Elbert
Thomas Elleman
Fred Erb
Captain T. W. Evans
Laurie Farmer
Nancy Farrow
Ed Faulkner
Maria Faust
Dave Fisher
Tom Fitzgerald
Howard Foley
Albert Freedman
Col. Funke
Gerry Furi
Mike Galle
Nancy Garcia
Susan Gately
Robert Gentile
Cecil Goodnight
Carolyn Gooley
James Gorman
Dennis Gormley
Rita Gotowala
Peggy Gottsacker
Jim Green
Ronald Greenwood
Don Hadley
Norman Hajjar
Harvey Halberstadt
Jerry Hall
Jane Halpert
David Hammel
Barbara L. Harris
R. L. Hassell
Lowell Hattery
Thomas Hayes
Meade Helman
Art Hillman
Norman Himes
Nancy Hoisman
Doris Holzheimer
Robert Hopf
Max Hopper
Glen Housey
Warren Hoyt
Don Hutson
Evelyn Jacquez
Steven James
Glen Jeffrey
Bernie Johnson
Ed Jones

Lowell Kam
Joan Karimi
Al Katzen
Bruce Kelsey
Barry Kirk
Diane Kirik
Terry Knight
Vernon Koch
Joe Kolman
Kim Kostere
Bill Lamb
D. W. Lamb
Mark Lansberg
John Larney
Joan Lefkowitz
Dale LePage
Robert Levite
Henry Lewis
Bob Lipson
Herbert Longfellow
Joe Lomker
Homer Lovvorn
George Lumsden
Clint McGhee
Thomas D. MacLeod
James Madden
Dayton Males, Jr.
Dr. Ralph Margulis
Martin Marietta
Louis Martin
Michael Meade
Richard Metzer
Brian Michmerhuizen
Dr. Richard Mignerey
George Miles
H. "Skip" Miller
James Miller
Jack Mooney
Dan Moseley
Sue Mutty
Kay Myrch
Patricia Newman
Ralph Nichols
John Nocero
John Oakes
Harold O'Brien
Mike Oliansky
Vaughn Organ
Floyd Parks
Bud Parrish
Roger Peay
Jim Penfield
Mike Phipps
Bob Pikes
Jean Pizzi
Ronald Plue
Preston Pond
William Power
Edward Pritchard
Ronnie Prophet
Samuel Raguso
Jackie Ramieez
Robert Ravitz
Bob Reuber
Howell Ridley
John Riser
Ladonna Robson
Ray Rosbury
Selma Rossen
William Roskin
Mark Ross
Ronald Rossi
Ree Rost
William Ruse
Edward Rusin
Wendell Russell
Robert Sarr
Tom Scullen
Cindy Seager
Theodore Settle
Malcom Shaffner
Randy Sher
Donald Shirey
Paul Singer
Rick Sklar
Peter Slaynasky
Richard Slayton
Rodney Smith
Jerry Solomon
Greg Stein

Kay Stites
Timothy C. Stockert
Richard Swanson
Lloyd Swartz
Ronald Swift
Donna Sylvan
Larry Taylor
Dr. Jerry Thomas
Wayne Totin
John Townsend
Danny Vitra
Hugh Wachter
Joseph Waldron
Rick Walker
Charles Walsh
William Ward
Mary Ann Welden
Edward White
Lee Wilson
Tom Winninger
Maria Wojtczak
Ray Wolfe
Gary Wolford
Robert Wood
Elizabeth Woolfe
Walter Work
Mary Young
Henry Zanardelle
Dr. Ziegler
Orin Zimmerman
Rita Zimmerman

A very special thanks to:

Bill Allen
Hector Aponte
Mohammad Aslam
Joseph Blomker
Ron Burke
Jerry Cordrey
Robert Davis
Glenn Jeffry
Jeffrey Justice
Joan Lefkowitz
Howard Lindberg
Michael M. Meade
Sue Mutty
Karen O'Kain
Kevin O'Sullivan
David Sansbury
Theodore J. Settle
Gerard VanderLven
Walt Work
Kathy Zanbetti

All the people at the Farmington Library.

The people at Gould, Inc., Computer Systems Division.

My staff of Dora Cole and Sandra Onka and a group of exceptional performers at Random House: Paul Donnelly, Linda Goldfarb, Patricia Haskell, Jane Rathbun, Helen Louie-Sumser and Matthew Thompson.

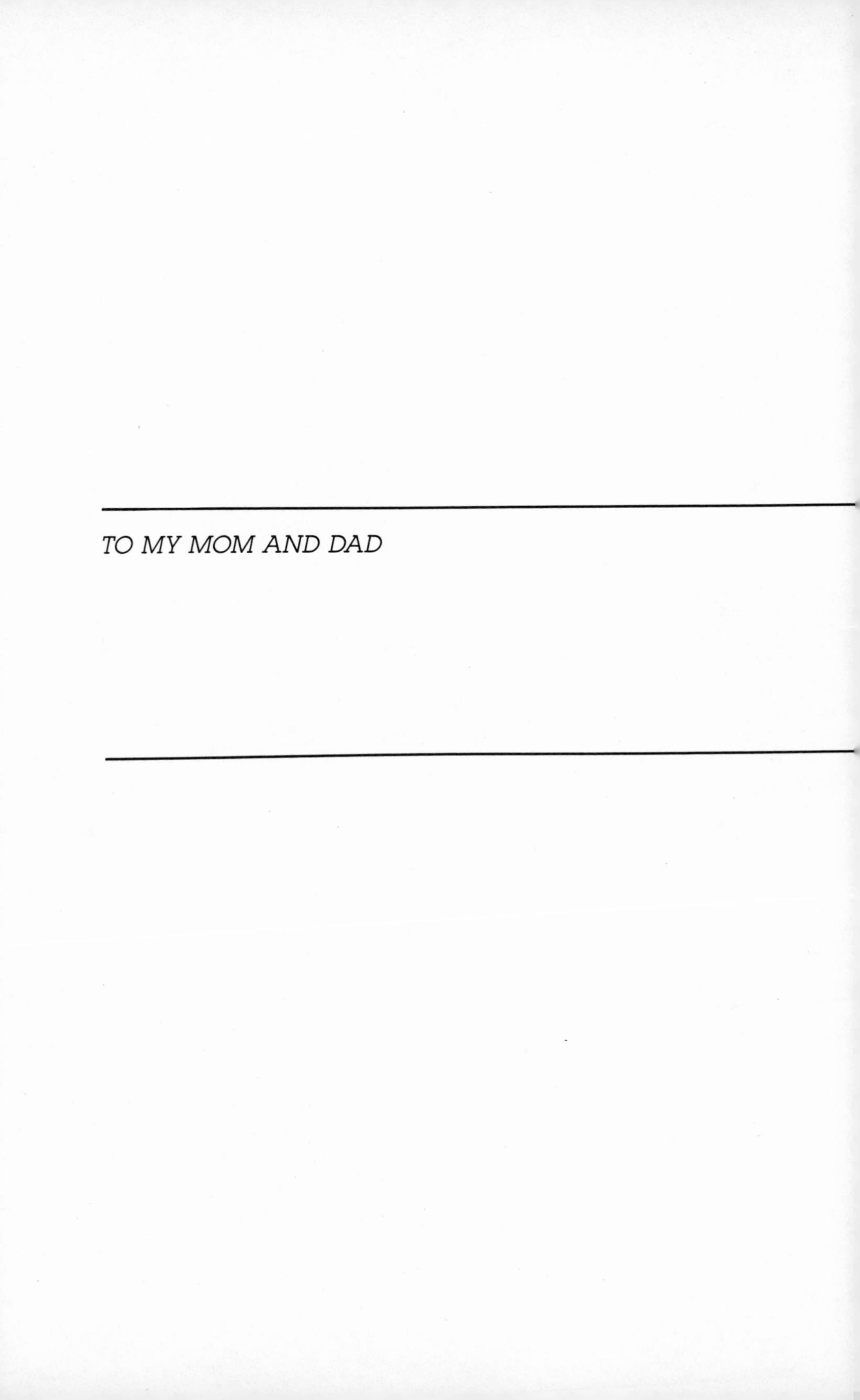

TO MY MOM AND DAD

WHAT THIS BOOK IS ALL ABOUT 1

Relentless natural selection in the modern jungle of business practices has quietly produced a new species of organizational manager.

Early in the '80s, some of the characteristics of high-technology management began to receive informal attention in the management world. In the casual talk after board meetings, in the restaurants where many marketing actions begin, and in jets over Detroit and New York, Tokyo and Bonn, managers began to sense the emergence of something different.

High-tech management was first thought of as a new style. As time passed, however, it appeared that more than style was involved. Perhaps there was a true substance to the successful management of high-tech companies that might deserve formal attention. Was a new cadre of managers developing, mainly concentrated in high-tech companies, that was doing something different? Something better?

One thing was certain. Nearly everyone *thought* that high-tech managers were different. The high-tech managers themselves thought they were different. And it was inescapable: they thought they were better.

My business is consulting with and training managers. If there had arisen a new cadre of managers with an image and tools that made them better, I needed to know about it. I made a major commitment to tackle the deceptively simple questions:

- Are high-tech managers different from managers in more traditional industries?
- Are high-tech managers better?
- If they are either different or better, how?

Now, two years after that commitment, I report the results of my investigation. The findings have implications for many management situations and certainly for many managers' approaches to the job. (See the Appendix for a discussion of the methodology used in the National High-Tech Management Survey.)

Two distinct concepts emerged:

1. High-tech managers are different because they perceive themselves to be different.
2. High-tech managers emphasize freeing themselves and others in the organization to use their enthusiasm, responsiveness, flexibility, and creativity to maintain a high level of purposeful activity. This emphasis augments but does not replace a more traditional concern with maintaining the formal structure and cohesion of the organization.

Both of these ideas, discussed at much greater length in chapters 2 and 3, provide the conceptual foundation of this entire book. Let's look briefly at the two findings now.

In a word, high-tech managers see themselves as better managers. It doesn't matter for the moment where this belief originated or whether it was well founded to begin with.

What does matter is that such a self-image tends to function as a self-fulfilling prophecy. A manager who sincerely believes himself or herself to be more capable, more creative, more able to overcome myriad obstacles and achieve substantial goals is likely to become so. These managers who see themselves as different and better right off the mark set higher expectations for themselves and their work groups. They find exceptional energy and drive for performance at a higher-than-routine level.

Further, these managers' heightened expectations appear to be supported by a unique twist on their relations with other people in the organization. Their view of others seems to derive more from what they have themselves learned from human interactions than from the older, production-oriented management principles. It is as if they bring to the job the high-tension mutual support and intuitive interresponsiveness found on sports fields the day of the big game. They contribute, and expect from others, the deep-seated personal involvement formerly found only in emotion-packed situations such as battle, sports, and dangerous exploration.

This involvement revives some principles of leadership and inspiration that have long been honored in word but neglected in deed. The exceptional performing high-tech managers give their subordinates, peers, and even superiors a sense of self-esteem, personal mastery, and self-determination. Specifically, they have sixteen managerial characteristics and accomplishments:

1. providing experiences to learn from
2. practicing letting go
3. using symbols and slogans
4. creating the excitement to achieve
5. identifying the expectations of the organization
6. making expectations explicit
7. building the expectations package
8. developing and communicating a broad vision
9. focusing sharply on what is important
10. acting decisively to empower others
11. communicating a sense of competitive urgency
12. creating activity and momentum
13. sensing the situation
14. sharing the glory
15. targeting performance
16. maintaining sang-froid

I hope to show that the genuine application of these sixteen factors, within a general organizational environment of heartfelt commitment and personal involvement, can produce results that justify the high-tech managers' view of themselves as different and better. I plan, further, to make these new management priorities concrete and accessible enough so that any manager can adapt them for his or her own area of responsibility.

Facing the uniquely harsh business environment of high-technology enterprise, high-tech managers survive because they apply attitudes and skills that permit long-term, virtually permanent, exceptional performance by management and technical development teams. In most high-technology industries a company cannot survive and prosper by producing routine, "reasonable" performance. Outstanding management achievements are needed just to stay in the game. It's life in the fast lane.

So where does that leave the rest of us?

In that same lane, whether we like it or not. The reality is that the entire business world is coming to resemble the threatening environment of high-tech industry. Changes in the American economy and the pressures of global competition are rushing the entire corps of professional managers, in all industries, into a world where this continuing exceptional performance will be needed for success. Increasingly, *all* industries are demanding the kind of exceptional management performance that the high-tech companies have required for survival; *all* managers must soon begin to develop some fast driving skills or find themselves in flames on the median strip.

This is a far cry from the routine filling in of process forms and the routine processing of management controls. The crackling excitement in the new project teams—whether in bioengineering or consumer marketing—is a totally new experience for many managers who have never been part of such an undertaking. The new management spirit is challenging and, to some, intimidating. But it has unique rewards in satisfying human commitment and exceptional accomplishment.

Bill Power, vice president, general manager at Young and Rubicam, Detroit, identified this challenge when he said, "Today's managers have to use resources other than people. Senior ranks have to manage the complexity of the new technology and they're also being driven to become a part of this new technology—partly because our young managers are demanding to use the technology and we are willing to try."

This book, then, identifies some of the human and organizational skills that outstanding high-technology performers use to draw their companies above the crowd of also-rans. It explains the style and substance typical of the successful new-technology approach in such a way that *any* manager can begin to adopt them and grasp the opportunities for success.

This book is also partly a "communications bridge," a way for managers who have succeeded in the high-tech environment to share their perceptions of the management job. It is pragmatic because it directly reflects the pragmatic styles of the managers who have contributed so much to its content. Principles and policies are not neglected. However, in every case emphasis is on what a manager can *do*. With few exceptions, if an idea or viewpoint did not suggest any practical applications to the seasoned managers I interviewed, I have left the idea out of the discussion.

How should you read this book? It depends on what you're looking for and where you stand in relation to the techniques and attitudes discussed. Based on your particular organizational environment and the way you conduct your daily work, you may find that some of these techniques and attitudes must be adjusted in their application or that some are more useful than others.

The major practical topics, the application of the sixteen essential management practices that have emerged in their full power in high-technology industries, are arranged for easy retrieval. Key ideas are highlighted and summarized so that you can get a useful overview of each topic in only a few minutes.

At the same time, every topic is given a fuller treatment in a traditional narrative discussion. Full discussion of reasons and results is still essential for real understanding and particularly for bringing changes in attitudes. Slogans and symbols are valuable but a little more digging is usually needed to produce lasting change. Chapters 6 through 9 present these sixteen critical topics.

Surrounding these key points are several chapters that expand on the setting and significance of the management practices. Chapters 2 and 3 explore more of the meaning of high technology. They look deeper into the often-overlooked myriad effects of the new technology in even the most traditional manufacturing and service industries.

Chapter 4 explores more fully the characteristics of exceptional performance. It shows how the management practices identified by successful high-tech managers work together to create a climate that allows long-term outstanding achievement.

Chapter 10 is in a sense the most important. It first develops a structural framework through which you can make the logical transition from the experiences described in the book to the management responsibilities that directly concern you. Then it proposes a way that you can develop a fairly systematic plan of action to manage your future involvement with the high-tech environment, however it impinges on your management job.

No matter in what industry, public organization, or branch of government you plan to develop your career, you are nearly certain to benefit from the experiences and attitudes of the exceptional performers in the survey. The concluding chapter is designed to ease the process through which you develop your own concrete benefits.

DEFINING THE "HIGH-TECH" ENVIRONMENT 2

Up to this point I have been using only the loosest notion of "high technology." Giving this key term a precise working definition is an important goal of the next chapter. Let's try now, though, to better understand the kind of environment in which high-tech managers have to work. This will provide a foundation from which to explore the potential effects of the new technology on managers in traditionally stable American industries.

In a stable industry, products and manufacturing and marketing processes tend to have a long life span. Consumers' buying habits typically are well developed. Competitors have no choice but to concentrate on fairly minor erosions of market share because scientific or engineering inventions are, or seem to be, unavailable to revolutionize product performance and thus market penetration. It is typically accepted that revolutionary change in production or marketing processes is equally unavailable.

Conditions are exactly opposite in high-tech industries. The pace of product development and improvement is so rapid that some life spans have literally been measured in months. This rapid development appears to result from both the relative ease with which performance improvements can be achieved and the adoption of aggressive product research and development as a part of the overall business culture in high-tech industries.

Since the high-tech industries' products are often improved through genuine technical pioneering, oftentimes the newer products cannot be produced by using existing plant and manufacturing processes. New high-tech processes often are not just different designs to be produced by the same old milling machines and injection molding stations. It is not unusual for a new product to require a brand-new manufacturing process.

The radical differences among succeeding generations of high-tech products or processes bring about a changed competitive picture. Advances or losses in market share are not so evolutionary as in traditional industries. Market shares do not simply rise and fall by tiny percentages as product features are gradually changed or marketing procedures fine-tuned. A competitor's new product or market strategy may be an order of magnitude better than our current best. Product, manufacturing, and marketing coups in many high-tech industries may put the competition out of business, not just steal a marginal market share.

In short, high-tech industries present an environment of continual change. The nature of the change is such that mere persistence and hard work are not adequate responses; creativity and truly exceptional performance are demanded. Further, miscalculations and the adoption of faulty strategies have more serious consequences than in more traditional settings. The survival of the company as a market force often may be at stake. It's not just a matter of a small error in judgement costing you, for example, a loss of 3% share of the market; it might well result in the company's total failure.

THE GROWING PERVASIVENESS OF HIGH-TECH

There's no denying that the introduction of technology creates as well as solves problems. Bill Creighton, C.M., Hospital Administrator at Blanchard Valley Hospital in Findlay, Ohio, addressed some of the problems technology has caused in hospitals. For example, they are having a more difficult time staying on the leading edge, especially when Medicare Disease Related Groups (DRGs) are expanding.

> "On a day-to-day basis the introduction of a new piece of technical equipment has a tremendous impact. It creates the need for training and some re-training, introduction of new policies and procedures, and the influx of new physicians. The sophistication of today's technology has also affected a manager's ability to shift from one specialty area to another. For instance, a manager in respiratory therapy can't as easily switch to surgery, because he or she won't be able to grasp all that has been brought out in laser [technology]."

Echoing and expanding upon what Creighton discussed is Ralph Margulis, obstetrician, gynecologist, and bank board director. He believes that technology has contributed to the changing times in the medical profession at large.

> "There are three factors operating today in the medical profession. First, advancements in technology, in combination with other factors, have caused hospitals to encroach upon physicians. Second, the doctors are able to capitalize on the technology more easily and develop their own facilities. They can use the technology of their choice at a lower cost, thus becoming more independent from the hospital. The third factor is that business and industry, through new technology and both physician and hospital dependence on third-party payers, are dictating to both hospitals and physicians to become more involved in rendering health care through HMOs [health maintenance organizations] and PPOs [preferred provider organizations]."

A reasonable response to the difficult characteristics of the high-tech environment might be, "So what! Who wants to work in that world? Let them have it!"

This response may not be a possibility for long, however. Several factors are conspiring to push every manager into the high-tech business environment.

1. Most companies of any size have internal departments or divisions that already are part of the high-tech culture. This can be seen dramatically in industries such as insurance and banking. Companies in these traditional industries cannot compete today without the service of the most

advanced electronic information-handling capabilities. Managers in the more traditional work of the company will be at a severe disadvantage without some understanding of and facility in the characteristic management styles of high-tech departments.

2. This derives from the first factor. Even if your company does not deal in high-technology products and markets, it is almost certain that increasingly the *use* of high technology will become essential to your job and your department. Eventually, companies that make a pervasive use of technology throughout their management structure and operating facilities come to exactly resemble companies in the commonly accepted high-tech fields of electronics, robotics, and information science.
3. Foreign competition, decreasing profit margins, increased transportation and other costs are all building to a level at which routine performance will no longer be adequate in any industry. This appears to be true even in industries that have been almost totally traditional and stable, with long-lived products and production processes. Throughout much of the American industrial base it is becoming increasingly hard to make a buck. The companies that do make profits year after year will be those with managers who achieve the exceptional performance of many managers in high-tech industries. Companies without such managers will inevitably fall by the wayside.

There is a sense in which the term "high technology" describes merely the relative newness or effectiveness of cultural products or methods for making these products. Thus, when our distant ancestors were carrying stone axes, a neighboring clan was a high-tech elite if they had managed to trade for three bronze scrapers and a knife. To a man with a sundial, an hourglass is high tech; to a man with an hourglass, a pendulum-driven clock is high tech.

So at any point in history there will be leading-edge applications of the current scientific understanding of the world; these may properly be called the high technology for that period. In this view there is nothing unique about current high-technology enterprises in information management, genetic engineering, robotics, organizational analysis and management science, or any of the many other current leading-edge applications.

"High-tech" is often used with this meaning in management discussions. Current engineering applications of the most recent scientific discoveries, whether in end products or in production processes, are typically and legitimately referred to as high tech.

This is not the sense in which I propose to use the term. I have come to believe that leading-edge technology is exploited in such ways, on sufficient scale, and with such a pace of inventiveness that a unique high-tech management segment has developed among practitioners of such engineering and scientific magic.

THE NEW NEW WORLD

A notable feature of the new world of high-tech management is the nearly universal acceptance of the proposition that long-term exceptional technical and management performance is normal and expected behavior. A level of performance that would be highly unusual in many other human activities is viewed as standard here. A promising computer company goes under and the CEO loses both his job and his considerable capital investment. The industry response is "He just couldn't quite turn out what was needed as a manager." Yet that manager's skills and output were of a high quality and quantity that would be almost unheard of in many other fields.

The high-tech culture supports the cultivation of a highly charged atmosphere of intense personal involvement with the goals and activities of the organization as a means of making exceptional performance possible. Many of the social beliefs supporting this involvement have now become so widely accepted that they can be identified as new "principles" of management. They are not true principles, perhaps, because managers do not learn them in management training or academic study and then apply them on the job. Rather, the managers appear to adopt the practices because that's what comes naturally; it is only on later reflection that the elements may be separated and expressed as abstract management principles.

We are trying essentially to discover these practices and to find how they fit together to produce exceptional performance. In this chapter, a central goal is to identify and explore the high-tech environment within which the new management practices have gained their first wide currency.

WHAT DOES HIGH-TECH MEAN?

"High-tech" has purposely been used loosely up to this point in our discussion. It is time for a more rigorous definition.

"High-tech" commonly refers simply to computer hardware and software. People who use computers may by virtue of this use refer to their jobs as high-tech. For example, employees and managers in the data-processing subsidiary of Sunbank in Florida are computer users. They are fiercely proud of their position within their bank as the high-tech people, a new-wave unique corps within the organization.

These employees have adopted a distinctive mode of dress; they have an unstructured working environment with flexible working hours. They have won salaries somewhat higher than the prevailing levels in the bank. They feel conscious pride in their position and view their work in the organization as on the cutting edge of the new technology.

Alternatively, high-tech means direct involvement with scientific research or with the engineering application of very recent scientific discovery. This orientation stresses the *creation* rather than the use of technology. Computer and systems design, genetic engineering, the development of electronic instrumentation or numerical control systems, work in semiconductors or exotic energy sources are the sorts of enterprises that come to mind under this emphasis.

I have often observed the concrete reflections of the scientific project culture when I have provided training at Los Alamos Laboratories in New Mexico. Some of the scientists and technicians there burn with the zeal to produce. They find it difficult to relax; they constantly talk about their current project. More often than not, they work very long hours without interruption and eat only occasionally and then skimpily. Some have even installed cots in their offices so they don't have to waste time commuting.

Both of these concepts—technology as used to support other goals and technology as the direct subject of work and creativity—are legitimate aspects of high technology. Both users and creators view themselves as different and unique. Both groups may be faced with the demand for constant inventiveness and with exceptional requirements for dealing with change and risk. Both must call on an unusually large, sophisticated body of technical knowledge in the "routine" performance of work.

To further analyze the characteristics of these two groups, I used information provided by the managers in the National High-Tech Management Survey to develop two scales measuring inventiveness and efficiency. The inventiveness scale is meant to reflect the extent to which a company or industry accepts the goal of continually implementing innovative technology, even in the absence of an immediate market requirement to do so. The efficiency scale provides a means of describing the extent to which technology is accepted as a primary means for solving internal organizational problems and the effectiveness with which technology is applied for this purpose.

THE INVENTIVENESS SCALE

The adoption of inventiveness as the central measure was dictated by the interview responses in my survey. According to this view, the extent to which a company or department reflects the high-tech culture depends most importantly on the perceived value of innovation for its own sake. The specific products of the company or department, of course, contribute to the adoption of inventiveness as a business strategy. Some industries and markets have been shaped from their beginning by competitiveness based on innovation. Individual firms in those industries and markets are limited in their freedom to adopt a different strategy.

In many cases, however, firms are not *required* to adopt the inventiveness strategy. True high-tech firms, according to this definition, engage vigorously in innovation and creation even when they are not required to by market demands for new and improved products or services.

Thus, any company might be rated according to its level of inventiveness. The scale applies as much to a pork processor as to an aerospace design firm.

The inventiveness scale consists simply of a vector ranging from lesser inventiveness to greater inventiveness. An arbitrary number of unlabeled midpoints may be defined between the least and greatest measured inventiveness. Points lying on the scale should be unnumbered to avoid the interpretation that greater inventiveness is necessarily better or higher. The purpose of the scale is to model states of nature, not to reflect a perceived value of those states. Six factors contribute to the overall position of a firm on the scale.

1. *Rapidity of technological change*. Inventiveness is obviously directly related to changes in both products and production processes. The extent to which individual firms elect to engage in these changes depends on both the industry of which they are part and the strategies they adopt within their industry. The construction of single-family frame houses, for example, has changed only slightly in the last half-century. Only very recently have significantly different materials and methods begun to be widely adopted. Home construction companies have felt little pressure either directly from the market or

from competitors to engage in technological change, although considerable change is possible. The second situation—the pursuit of different strategies within a rapidly changing industry—is well illustrated by the manufacture of wrist watches. Digital watches use a continually evolving array of semiconductors and exotic display devices. Yet a number of watch manufacturers, notably Concorde and Rolex, have maintained a very competitive position within their market segments while adopting virtually none of the new digital technology in their products.

2. *The threat of obsolescence, caused by either market demands or cost escalations*. In this case, the product line of a company or department dictates the need for inventiveness. Two situations apply. Competitors continually introduce new products whose improved performance will steal larger and larger market shares. A company in this market has no choice but to respond with its own new products or, eventually, to withdraw. Alternatively, competitors—domestic or foreign—continually develop more cost-effective manufacturing processes and management systems. These improvements reduce the final price to consumers, again yielding greater market shares and forcing response by other companies. Frequently a firm will be faced simultaneously by both of these threats: a continual flow of new products at continually reduced prices.
3. *The possession of exclusive knowledge*.

Typically such knowledge consists of traditional trade secrets related to internal material formulas and manufacturing procedures. Increasingly, however, market information, specialized skills of technical staff, or even close personal connections with sources of venture capital and entrepreneurial assistance have the same effect. Many firms in the information industries, for example, have turned a single excellent mailing list into an effective competitive tool. A software firm lives and dies by the individual skills and creative ideas of its analysts and programmers as well as by the effectiveness of unique control and testing procedures. A semiconductor firm gains precious competitive months through new precision control of chip production.

4. *Exclusive knowledge providing the potential for growth*. This depends mainly, of course, on whether the exclusive knowledge yields marketable products. It further requires, however, a continual development and updating of the knowledge itself. Typically such development derives from both conscious corporate strategy and the presence in the organization of people for whom knowledge is its own reward.
5. *Substantial budget for research and development.* There is no standard proportion of budget that inventive companies devote to R&D. The proportion does matter, though. One study showed that on the average the leading firms in an

industry devoted as much as one and one-half to two times the industry average investment to R&D (measured as a percent of sales). Further, the long-term leaders typically have the largest development budgets on an absolute dollar basis.

6. *Frequent major product announcements.* This is, of course, the concrete pay-off of many of the previous factors. It is nevertheless important in itself. All of the knowledge, technical development, and management expertise in the world will mean nothing to a company unless it successfully crosses the last hurdles of product introduction. Good technological ideas abound; it's only via developed, working products, however, that corporate returns can be obtained. There can be no doubt: among high-tech companies, the industry leaders are the product pioneers.

Product Adoption Stages

During an intensive interview, Bob Gentile, a marketing expert for a Boston computer firm, characterized the products of the most inventive firms: "From a marketing context, the 'newness' of new technology can be explained by referring to the adoption curve." This is the familiar bell curve in which total product sales are plotted against time from introduction. Brand-new products gain slow acceptance, and the slope of the curve is slight to begin with. At some point the slope begins to increase because total sales have grown to a point to justify price decreases or because the exceptional value of the new product has come to be recognized. The slope then becomes quite steep for a time and levels into a plateau. Product obsolescence or market changes then reduce sales volume fairly quickly until finally only a very small market remains.

The point, according to Gentile, is that high-tech companies, operating by choice or forced by circumstances, tend to maintain their product mix with items that are predominantly on the upslope of the adoption curve. Further, the adoption curves (or life cycles) of high-tech products tend to be short. For some products the maturity plateau may be reached in a matter of months.

THE EFFICIENCY SCALE

The inventiveness scale applies most directly to creators of new technology. The equivalent scale for users of technology is the efficiency scale. It models the level of adoption of technology for the internal operations of an organization. Seven factors contribute to this level in an individual company or department.

1. *Automation of routine tasks.* Most managers recognize that routine operating tasks can be automated to a considerable extent. The extent varies among industries and departments. The degree of commitment to undertake internal automation is a measure of the efficiency of the adoption of technology. Considerable commitment is required to achieve any significant level of automation. Repetitive, clearly definable tasks must be identified first. This typically requires a major analytical effort. The development of automated routines to replace existing procedures calls for a high level of ingenuity and skill in dealing with people and organizational systems. Then a further

major program is needed to install and work the errors out of new technology-based systems. To undertake this task wholeheartedly demands a dedication beyond that brought about by a simple cost-benefit analysis.

2. *Recent training in new technology.* A company that does not make the investment of continually updating employees' skills in the use of technology has not made a full commitment to efficient use. Constant change of approach is undesirable, according to most analysts, but relatively small adjustments and replacements as technology improves do pay off.
3. *Access to new information.* To maintain efficient use of technology, managers must know what technology is available. Training provides part of this knowledge, but training development times often cause true leading-edge conditions to be left out. Access to trade shows, personal industry contacts, periodicals, and product literature indicates a desire to maintain efficient use.
4. *Effectiveness of overall information system.* Efficient use of technology thrives on the rapid, widespread internal dispersion of a wealth of information. The companies that are best able to make use of technology will have the apparatus in place for identifying and collecting a large body of relevant information. This information will be systematically sorted and then communicated to all management and technical levels of the company.
5. *Presence of technological "think tanks."*

The existence within the organization of a group or groups of people with the express goal of planning and creating future uses of technology is nearly essential for efficiency. A large organization is too complicated for haphazard applications to work. The sensitive interactions of different technologies combined with the delicacy of some of the problems of user acceptance of new methods demand careful planning at a central point of responsibility.

6. *Auditing for full utilization*. Efficient users periodically evaluate or audit the total application *system* for the use of technology. Only in this way is it possible to detect and react to imbalances, gaps, and obsolescence.
7. *Adding value to new technology*. Even the user does not have the choice of simply adopting technology. The user must tailor and create the specific applications that are most effective for the company's operations. In particular, this internal creation and improvement of applications is effective if, by policy and in fact, the guiding principle is increasing the competitiveness of the firm in its markets.

Clearly, a single company can be rated on both of these scales. Many companies score high on the factors for both the efficient use of technology and the inventive creation of technology. This need not be the case, however. Small firms in particular may succeed as creators but not have the management structure or resources to use effectively the technology available in the market. Many traditional or mass-consumer firms adopt a strict follow-the-leader policy in both products and production processes. These same companies, however, may be among the world's leaders in the efficient use of technology in internal activities and management systems.

WHAT MAKES THE HIGH-TECHS DIFFERENT FROM TRADITIONALS 3

Analysis of the information provided in the survey interviews suggests some ways in which high-tech companies—those high on inventiveness, efficient use, or both—differ organizationally from traditional companies. These organizational characteristics reflect the ways survey managers viewed their own companies. The perceptions do not necessarily relate to the formal policies of a given high-tech or traditional company.

This is how managers in high-tech and traditional firms described their own organizations on nine important factors. The descriptions are not, of course, absolutes. They reflect emphases. When a manager says that her company stresses the development and use of formal communications channels with important messages always conveyed in writing, this does not mean that there are no informal communications in her company.

1. Communication

   Traditional Formal. Messages are intended to flow through planned communications channels in routinized reporting procedures. The channels follow closely the structured chain of command, although lateral and diagonal channels may also be established.

   High-tech Informal. The origination and passing of information and decisions is not meant to be structured according to the company's internal chain of command. Expertise and interest in a particular problem are more important criteria for who is included in communications related to the problem. Personal contact and response to the demands of the instantaneous situation prevail. It isn't uncommon, for example, for a manager to "go over someone's head" in pursuit of problem solutions. Doing so is not officially frowned on as it typically is in traditional companies.

2. Status

   Traditional Symbol-oriented. Power is officially identified and derives importantly from the formal position a person holds. An ineffective marketing manager, for example, has almost exactly the same effective authority and receives the same deference (in public, at least) as a superbly effective marketing manager.

   High-tech Ability-oriented. Status comes from expertise and skill. The person with the answer and the appropriate creative idea is likely to be recognized and deferred to, no matter what his or her organizational position.

3. Climate

Traditional Stable. The prevailing stability of the traditional climate derives from a number of factors. Product life cycles are longer and the pace of development and introduction can be more leisurely. More diversified product lines are common; this reduces the negative impact if a single product fails in the market. Marketing deals with the performance of relatively well-known products in well-known markets. Time is available for fine-tuning marketing efforts to gain relatively small market shares. Managers are capable of making decisions with direct reference to what has worked and failed to work in the past. Since decisions usually relate to relatively small changes in a small part of the product line, promotional effort, or market segment, any single decision is unlikely to have an overwhelming impact on the organization or industry.

High-tech Changing unpredictably. The very basis of competition is quickly developing new products, marketing ploys, distribution channels, and manufacturing processes. Repeatedly dropping unknown products into unknown markets characterizes the high-tech environment.

4. Decision Making

Traditional Centralized. In traditional companies, efficiency of operations is more important in maintaining competitive advantage than is innovation. As a result, control of operations emerges as the prime management concern. Centralization of decision making contributes to efficiency.

High-tech Decentralized. Decisions tend to be made wherever necessary to promote creative solutions to problems. Further, decisions that make little reference to past experience and to empirical support from previous operating results may often be accepted. To some extent this "history-free" decision process is forced on high-tech companies because good market experience will be lacking for very innovative products. In another sense, though, this sort of ad hoc localized decision making reflects the trust in intuition that often characterizes high-tech firms.

5. Focus

Traditional Policy-oriented. A legitimate response to a question about a decision is "That's the way we do things around here." Tradition, encompassed in written and unwritten policies, is effective at maintaining an organization that does not have to respond to a great deal of change.

High-tech Product-oriented. A division exists here. In fact, internal emphasis is reported to be on the products to be sold. Product characteristics tend to be derived from feasibility of design and production. This emphasis lessens the degree to which the needs of buyers in the market influence decision making. Yet respondents report a continual orientation toward customer satisfaction among high-tech managers. It appears that today there is sufficient market demand for new products routinely to succeed, even when specific customer needs are sometimes subordinated to technical concerns during the design stage. Many expect this condition to be only temporary.

6. Resources

   Traditional Capital-intensive. Efficient mass production is the typical goal, and the typical means employed consist of expensive production facilities.

   High-tech Total resource-intensive. Human resources are of especially obvious value to a firm that lives or dies by innovation. This is true to such an extent that the human impact of decisions has gained proportionately, compared with profitability, as a decision-making criterion.

7. Authority

   Traditional Authoritarian. Directions govern actions. This remains essentially true even in traditional organizations that nominally accept the human-autonomy teachings of recent academic work in management. When efficiency and effective control are the first-priority requirements for competitive success, a good deal of authoritarian management is necessary.

   High-tech Empowering. In the high-tech world, competitive advantage derives most importantly from the creative contributions of highly motivated and personally involved people. The successful companies in that world are those that have genuinely distributed organizational power to everyone involved in projects and programs.

8. Reaction

   Traditional Rational. The collection of data over long periods of time provides measures and estimates for use in a decision-making process that is made as logical as possible.

   High-tech Intuitive. In the first place, the volumes of data needed for quantitative and logical approaches to problems are more likely to be unavailable. Further, there is inherently greater respect for the value of the intuitive leap in products and business practices. This intuitive response can and sometimes does take the form of unfocused "shooting from the hip." In the best cases, however, it may express uncommon abilities to synthesize nonrationally a tremendous number and range of facts and impressions relating to a decision situation.

9. Direction

   Traditional Backward-facing. Like it or not, the greatest opportunity for a company whose success is based on production and distribution efficiency is to be excellent at solving problems. Anything that disturbs the status quo of smooth operations must be quickly adjusted. Opportunities exist for achieving gain, but it is natural that emphasis comes to be placed on the opposite view: maximally avoiding loss.

High-tech Foward-facing. Always in the background of decisions is the conviction that some new product or way of doing things will blow the competition out of the water. Recognizing and grasping opportunities comes naturally when the atmosphere is buzzing with innovation. Carried to an extreme, projects or even current products that encounter problems may simply be dropped.

HIGH-TECH MANAGEMENT SUPPORTS COMPETITIVE ADVANTAGE

One conclusion is easily drawn from these factors identified by the survey managers. Companies tend to adopt the atmosphere, traditions, policies, and ranges of human relations that agree with the internal perception of the company's competitive advantages. Many of the attributes of the traditional firm—centralization, tight control, authoritarian management—may be seen as pejorative in today's world. Yet these approaches work in many industries. For these companies, only the highest efficiency in manufacturing and distributing masses of products will create profits. Whenever that is true, tight internal control becomes desirable.

The high-tech company faces a radically different situation. As the survey data show again and again, rapid change is the norm. The manager daily faces uncertainty, high risk, unpredictable market setbacks and opportunities, and unanticipatable competitive challenges. Individual and company track records are often short and inconsistent. They are typically of little help in decision making. Strategic planning comes to resemble an exercise in Zen meditation because so many factors are unknown and unknowable. The high-tech manager faces the proposition that it is impossible to exercise tight control over the unknown.

Flexibility under a barrage of new products and competitive approaches is essential for a company that chooses to compete in the innovative industries. This company flexibility can come only from the flexibility of its individual managers.

Cooperation among departments and work groups must be extreme; this requires managers to act at least as if they have almost unlimited trust in each other. The immediate goals of the marketing group and the manufacturing group will often be considerably at odds, for example. Marketing may want to guarantee a certain price to customers while manufacturing cannot yet be sure that costs can be kept low enough to make that price profitable. What if the manufacturing manager commits to the price anyway? It takes naïveté to think that the manufacturing manager won't be blamed for the failure if his department doesn't meet the low cost figures. Yet no one knows in advance whether they can be met.

It seems, then, that ragged-edge technology demands ragged-edge human relations. A "smart" business manager from the traditional mold would be first concerned with protection from failure and blame. The Hippocratic aphorism for doctors, "First, do no harm," has been revised wisely in many traditional companies: "First, do no harm . . . to me." This natural, protective attitude will not work, however, when exceptional creativity and performance are demanded throughout an organization.

The survey results bring out a final point about high-tech companies and approaches to business development: increasingly, all companies are moving toward the high-tech environment. Several reasons were mentioned for this direction.

1. The use of technology for internal control and management systems in traditional

companies is bringing increased general awareness of the rage for the new that typifies the leading-edge firms. Managers in these internal departments often bring a typical high-tech point of view to their work. One thing that makes high-tech managers innovative is their belief that innovativeness is the best way to run a company. This belief has proved infectious.

2. Numerous factors in the economy—notably, foreign competition, higher average charges for the use of money, increasing energy costs, and aging physical plant—are creating profit pressures for traditional industries. Profits from operations are shrinking in many sectors. It is easy to name companies that make as much money from financial transactions, such as currency exchange, as from the production business they were originally formed to pursue. This pressure has begun to lead managers in traditional industries to adopt some of the bold strategic and product-oriented thinking of the high-tech companies. If this style of management catches on in traditional industries, they won't remain traditional.
3. Traditional companies usually emphasize efficiency of operations. This central goal has been a factor in keeping traditional companies traditional; stable, highly controlled environments are essential to long-term efficiency. However, we are now entering a period when the very tools of efficiency—internal production equipment—begin to undergo very rapid

> technical advance. The technology revolution largely began in the realm of information creation and exchange. Its growth in the design and production of manufacturing tools is only beginning. The effect of the introduction of robotic equipment and advanced programmable manufacturing stations and modules will be to make many of the long-term stable companies need continual innovation as much as does today's rawest scientific invention firm.

The ultimate goal of exceptional management performance is to affect the organization in such a way that employees and managers get the competitive work done. The central feature of exceptional management performance is to affect *others*, not to do the work yourself. In this sense, exceptional performance might be called "pure" management. It is concerned *only* with allowing others to produce optimally. It aims to let others make decisions, be creative, analyze and accept risk, innovate, judge the quality of their work, and turn out an exceptional quantity of work that is directly related to the accomplishment of corporate goals.

Employing this sort of pure management contributes directly to building a high-tech enterprise. Many of the characteristics that work to make a successful high-tech company derive from the features of exceptional performance. Let's look in more detail at the sort of organization that managers try to build in the high-tech environment.

THE HIGH-TECH ORGANIZATION

According to survey respondents these are the chief characteristics of a **successful** high-tech organization:

- It applies nonroutine technical and commercial approaches.
- The approaches address challenging and rapidly changing production, marketing, and management situations.
- The working environment must be flexible to deal with the rapid changes.
- Results of operations are measurable.
- The creative element of operations, in particular, must be efficient in the use of financial and human resources.
- Problem identification and solution must be intuitive, rapid, and effective.
- Operations produce a new or expanded body of knowledge.
- The outcome of this new knowledge is the introduction to the marketplace of products or services that have both better quality and more cost competitiveness.

In discussion with scores of managers, I was struck by how consistently the key features of this definition of high-tech organizations were displayed. Even the largest companies in high-tech fields have a more informal and flexible working atmosphere than more traditional firms. Office doors are open, work hours are less rigid, dress is often casual. More importantly, direct communication among all layers of management and with nonmanagerial employees is typical. Neither the formal structure of the company nor the job descriptions of particular positions block the flow of information. The ideas of a recently hired junior technician will often be considered as important as the ideas of the CEO. These companies also have fewer guidelines, fewer established rules and procedures. This may be a consequence of the newness of many high-tech companies and departments. In any event, the lack of standard methods promotes flexibility.

Flexibility and fast, direct communication feed creativity. This environment has the effect of nourishing nonroutine approaches to

- product development,
- meeting deadlines,
- working with the unknowns inherent in technical development,
- managing creative people effectively and efficiently,
- responding to crises,
- cashing in on unexpected opportunities, and
- solving problems.

Inventiveness and the willingness to accept risk become highly desirable traits at all levels in such an organization. Beyond this, however, high-tech organizations need to allow employees and managers the chance to express their own human desires and interests in the workplace. At Apple Computer, for example, many employees believe they are making the world a better place through their work. It is essential that these high-level human interests become available as a motivating factor in the high-tech world. Exceptional performance is possible only when employees are "turned on" by the things that are truly important to them.

That brings us to the final group of characteristics of a high-tech department or company: its people believe they are doing some good. They use the new body of knowledge that their creativity has produced and *apply* it to placing products in the market that are truly better and more cost effective. The feeling of doing good for the world emerges again and again as an important motivating factor in high-tech enterprise.

Exceptional management performance is needed to create an organization with these characteristics. At all levels, a continuously high quality of performance is needed. Most people are capable of performance at this level only if they are passionately engaged in the vision of the corporation. High tech, as discussed above, has inherent qualities that lead employees to make a personal commitment and to produce their best, most inventive work.

INCREASING MANAGERIAL LEVERAGE THROUGH EXCEPTIONAL PERFORMANCE 4

Two characteristic results are achieved by a manager who performs exceptionally well. The first is that the people who report to an exceptional manager are allowed and influenced to do their best work. One manager surveyed said that people are not controlled—as they might be in a traditional environment—but facilitated. The exceptional performing manager provides ways for subordinates to use their commitment and interest to produce significant work.

The second result is greatly increased management leverage. Consider a hypothetical manager in a company that produces precision cameras for industrial use. The company wishes to develop and market a new bench camera. Say, for the purposes of discussion, that this manager has the technical skill to do all stages of product development, testing, production, and marketing. Alone, he might set out on this project in 1985 and finish somewhere around 1995.

Instead, the manager chooses to do none of the work himself but hires ten specialists in design, production, and marketing. When properly managed, these ten will produce more than ten times the manager's work each month. The work will be done better and far faster.

This is the situation the exceptional manager strives for. *All* of the technical and professional work will be done by others. The manager, by using his skills to motivate and facilitate others, is achieving leverage. His own skills are far more effective in management than if he used part or all of the time doing direct work.

The reverse is also true. A manager's poor performance is also leveraged. One ineffective manager reduces the level of output of all his or her direct subordinates. These subordinates then reduce the effectiveness of *their* subordinates, and so on down the line.

The most interesting result of the National High-Tech Management Survey is that exceptional performers can identify the behaviors they use to keep their own work and the work of their people at a consistently high level. From the study, four major performance areas emerged:

1. Building confidence, pride, and commitment
2. Managing expectations
3. Creating and managing events
4. Managing operating processes

These four major areas of concern work together to create an environment in which employees and managers can free themselves to do the most good.

The survey managers further identified a number of concrete behaviors that contribute to each of the four areas of concern. I have selected the sixteen practices that were most commonly mentioned and that created the most integrated approach to the management job. The remaining chapters of this book discuss these practices in some detail. The remainder of the current chapter shows how the four major areas of management concern and the sixteen management practices work together to produce a unified approach to exceptional management performance.

BUILDING CONFIDENCE, PRIDE, AND COMMITMENT

Exceptional performance can exist only within a setting that frees people psychologically to do their best work. Resentment, dissatisfaction, dislike of the organization's management style, poor training in the specifics of the job, a sense of isolation caused by poor communications practices, and myriad other factors arise frequently to interfere with work effectiveness. In the high-tech environment these human responses can be especially devastating, because the first element of performance to be damaged is the high-energy creativity upon which technological development depends.

Gary Wolford, Personnel Development Manager for Marathon Oil Company, is responsible for the administration of the company's worldwide organizational development and human resource management activities. In our first interview, I quickly discovered that Dr. Wolford and his staff espouse the need to focus on human tools in addition to technical tools for exceptional performance in the high-tech petroleum industry. In an effort to make his point, Wolford is quick to remind managers of oil-related incidents (not

Marathon-related) when poor management of people has had serious consequences.

A classic episode was the time an offshore oil rig was found to be missing some electronic equipment. It was discovered that the equipment was missing because an employee had pushed it off the rig into the water. The employee felt that management had treated him unfairly and he was intent on getting even. Or there's the transport driver who put sugar in the engine of a diesel tractor to show his frustration. For years he had been a good employee, always working a 7:00 to 3:30 schedule. Suddenly, management strictly enforced an 8:00 to 4:30 work-day schedule. Pouring sugar into the tank was his way of showing disgust; his wife was dying of cancer and had to go for chemotherapy each day before 4:00 p.m.

Wolford and his people emphasize the belief that "it's people that make the technology of their industry work."

The manager's job is made more difficult because many of these negative feelings appear to arise rather naturally in large formal organizations such as corporations. The manager has to work actively to sponsor psychological satisfaction; it is not sufficient merely to avoid mistakes.

The three factors exceptional managers identified to provide the proper psychological climate for best performance can be briefly described. These are the background factors that motivate employees to take risks and start activities to support the organizational goals.

- Confidence

Confidence derives from the conviction that we know how to do the jobs we are assigned and that the organization will provide the support and resources needed to help us perform well. Obviously, a manager can build confidence by seeing that subordinates are properly developed to have the skills and attitudes they need. Further, it is the manager's job to provide the support and resources subordinates need to perform.

- Pride

Confidence precedes action and makes it occur. Pride follows the action and stems from the knowledge that we have produced our best work. Accordingly, it is possible to feel pride even when our efforts have not succeeded. Pride stems from excellent performance and from knowing that we have accepted responsibility.

- Commitment

Commitment refers to the unstinting acceptance of the welfare of the organization as an important personal goal. If the company does well, we are doing well because the company is achieving something that is personally important to us.

MANAGEMENT PRACTICES

A package of four management practices was identified as helping build confidence, pride, and commitment:

- Provide experiences to learn from. We need to take responsibility for providing employees with the information, skills, knowledge, and attitudes they need to perform assigned work. We can do this through training and development, and by taking care when delegating work.

- Practice letting go. This is an extension of traditional delegation. We must let subordinates really carry the ball. They need to have the freedom to establish the conditions of the work they do. They also need to get the credit they deserve for work they have done. The distinction from delegation is one of emphasis: in letting go, subordinates are not seen as helpers of their boss. The subordinate is seen to have the *prime* responsibility for carrying out the assigned work.
- Use symbols and slogans. The use of shorthand indications of the vision of the company helps to build cohesion in work groups. If properly created and used, symbols and slogans will strongly reinforce employees' feelings of accomplishing personal goals through the medium of the corporation.
- Create the excitement to achieve. Build generous recognition of successful work into the system. The recognition should be suited to the personal style of the employee. Recognition perceived by employees as routine and obligatory will not work. Further, allow individuals to feel that their efforts genuinely contribute to the organization, show how the parts work together, and emphasize the importance of individual accomplishments.

MANAGING EXPECTATIONS

Confidence, pride, and commitment may legitimately be viewed as the background or atmospheric setting of exceptional performance. The three terms refer to the general way individuals relate to the organization and the organization relates to individual employees and managers.

The methods of managing expectations bring us closer to the performance of immediate tasks that make up a position in the organization. Operating within the general climate of confidence, pride, and commitment, individuals will have specific ideas about what they want from the organization, from their subordinates, and from themselves. Further, these expectations will refer to matters such as the acceptable level of quality, the amount of work effort that is reasonable, and the ways of working together that will prevail in the organization.

All of these expectations may be either spoken and clearly communicated or unspoken and communicated only in subtle ways if at all.

A prime example of conflicting expectations is in the hospital environment, as Dr. Harvey Halberstat, Medical Director of Woodland Hills Center in Troy, Michigan says:

> "A doctor operating in a nonprofit hospital is placed in a nonmanagerial position. The doctor is answerable to the patient, where the nursing and hospital staff is answerable to the administrator of the hospital.
>
> "The situation becomes even more complex with the onset of more high medical technology. The doctor continues to be concerned about the patient's well-being while the hospital administration is more involved in profit-loss concern in a growing competitive market."

When expectations are not congruent, as this example and others mentioned in interviews illustrate, operations management becomes more complex.

The manager's job is to see that the expectations from different levels of the organization are congruent. For example, if a manager wants to delegate the preparation of a report to one of the people who works for her, she knows what her expectations are in regard to the report. She knows the role the report will play in overall organizational efforts. And she knows what her personal expectations are about working with subordinates. The person being delegated to knows the conditions of work that are most comfortable for him. He knows what kinds of rewards he prefers. He knows the kinds of resources and staff he needs to work effectively.

In managing expectations, the manager's first task is to make all of the unspoken expectations as explicit as possible. There then follows a discussion and negotiation stage in which both sides try to find optimum compromises.

MANAGEMENT PRACTICES

The exceptional performing survey managers identified three essential practices that help to bring expectations at different levels into harmony.

- Identify the expectations of the organization. You cannot make employee expectations harmonious with organizational expectations until you get a clear picture of the types of organizational expectations operating. This includes the true vision pursued by the organization and the methods of working together that are supported by top management.
- Make expectations explicit. In a sense, this practice refers simply to influencing everyone involved to lay their cards on the table. It appears to be a natural human

trait to keep our most personal—and important—expectations to ourselves. This tendency can often be overcome by directed discussion. The manager can begin with the explicit requirements of the organization or of the work itself. Through careful questioning—of self and others—the less explicit and well-understood expectations can often be made to emerge.

- Build the expectations package. This practice consists mainly of devising specific events and ways of working that will satisfy the important expectations of everyone involved. It is possible, for example, that an employee's strongest expectations can be satisfied by allowing the employee to create the conditions of his or her work. Certain other expectations may be satisfied by monetary benefits or through public recognition of excellent performance.

CREATING AND MANAGING EVENTS

Successful managers do not manage people; they manage events that indirectly influence people. Similarly, managers do not motivate people; they manage events that create the conditions under which individuals motivate themselves.

Exceptional managers create events that involve the people in the organization. An event is a situation that a manager establishes with the aim of influencing others in the pursuit of important goals. The key point about effective events is that they be open-ended enough for others to create conditions under which they will work for the goals and volunteer their services.

At the point where a manager begins to create events, the manager's people should already be prepared to be highly engaged with corporate goals. They should have confidence that they can perform excellently under the right circumstances. They will be essentially committed to the organization's vision. They will have resolved their expectations at least in general terms with other levels of the organization.

In the management of events the manager presents his or her people with the opportunity to devote themselves to a specific task or project. The employees are given the chance to adopt specific tasks and work. The manager does not determine the exact procedures or orientation to use to accomplish the goals presented. Instead the manager describes the goals of the work, its importance to the organization, and the general sorts of resources available for accomplishing the work. The subordinate should be inspired and encouraged during the discussion to mention his or her personal wants, needs, and interests that are important to accepting the challenge of getting the work done.

MANAGEMENT PRACTICES

The exceptional managers suggested methods of managing events. These five were most often mentioned:

- Communicate a broad vision. People will not produce exceptional performance unless they are voluntarily working toward something they personally consider worthwhile. For this reason, it is not adequate for a manager to present only specific, numerical goals meant to be achieved by employees. Given a goal of reducing waste material by seven percent, a reasonable response is "Why?" If, however, the manager can communicate a broad vision of what the corporation is trying to accomplish, the reduction of waste can be seen as an important step toward a greater goal.

- Focus sharply on what is important. This is obviously a good idea because it helps in completing the true priority tasks without undue interference from extraneous or merely administrative issues. Such a focus serves a second purpose, however: most people find it more satisfying to give their attention to important matters. Few people feel a sense of fulfillment if their job is mainly made up of administrative tasks that contribute little to achieving the vision of the company.
- Act decisively to empower others. This does not mean merely to delegate. It means genuinely to give others free rein to accomplish tasks that move the company closer to its vision. Employees must seek their own best way to accomplish the work. If the work is completed successfully, the person who actually accomplished it gets the credit. The person doing the work is in direct control of the resources needed.
- Communicate a sense of competitive urgency. One thing that makes work important is its contribution to the company's ability to compete with other suppliers in the market. It has been a

tradition that the issue of competition is considered strictly the province of top management. The exceptional managers, however, believe that the urgency of success in the market is a concern of every employee. If employees and managers are going to make a personal commitment to the firm's success, they need thorough, accurate, up-to-date information about the status of competitors.

- Create activity and momentum. Initiative is central to exceptional management. As a rule, committed and confident subordinates, working in an atmosphere that satisfies their expectations, will generate the major initiatives. Times will arise, though, when the manager must take decisive action and make things happen.

MANAGING OPERATING PROCESSES

Events are usually short-term actions or situations. They have a distinct beginning and end. They are observable as creations intended to influence operations in a certain way. Processes are the continuing activities that connect events and make them possible. If negotiating a budget is an event, for example, the efforts of the employee and his boss to prepare for the budget meeting are part of the process of the department.

The process includes all of the actions and work that contribute to the achievement of events. Dealing with this company operating process is as close to a traditional management principles approach as the exceptional managers in the survey ever come. They viewed this process as the "glue" that holds the structure together to support events.

MANAGEMENT PRACTICES

The exceptional managers proposed the following approaches as most important to keeping the process accomplishing its goals.

- Sense the situation. The effective manager must personally keep in touch with the workplace. It is critical that he or she spend time around the people who are doing the work. Managers must really have their radar on and think laterally and associatively to sense the true state of progress toward goals. This intuitive sensing must range over all sources of information: employees in the workplace, talks with other managers inside and outside the organization, messages coming from above and below in the organization, reading, watching TV, and so forth.
- Share the glory. This becomes natural in events because one purpose of events is to get the employee successfully engaged on his or her own. It becomes nearly impossible to *avoid* sharing the glory since it will be so easy to see who was really responsible for the success. However, people who contribute to the process—as contrasted with events—should also receive the glory for success. This often takes more conscious effort.
- Target performance. Sharing the glory provides positive reinforcement for outstanding work. Targeting performance refers to giving oblique feedback to reduce the extent to which employees create ineffective or poorly aimed per-

formance. Giving criticism is one of the hardest management tasks. It should never be overlooked, however; employees can maintain exceptional performance only if they know which work is good and which less good.

- Maintain sang-froid. Risk, crises, and problems are inevitable in any operating environment. The leader—the manager—must provide a positive model of coolness under pressure. Several techniques are available to make this calm response easier to maintain.

THE ROAD TO EXCEPTIONAL PERFORMANCE: THE FAST LANE AT INTEL 5

Let's take a moment for a story that points to some of the emerging factors in the business environment that call for exceptional management performance. The development and introduction of a new product at Intel Corporation show the exceptional organizational responsiveness, confidence in the management team, innovativeness, and plain guts that are typical of the new style.

THE FAST LANE AT INTEL

Intel could be considered the backbone of the Silicon Valley information hardware industry. In the company's Santa Clara offices, shirt-sleeved engineers roll out innovation after innovation in the design and manufacture of semiconductors. These are the intricate building blocks of modern electronic equipment from computers to arcade games to solar-powered wrist watches to electronic auto ignitions. The devices are called "chips"; collectively the many different kinds of semiconductor chips have been the central element in development of the modern information industries.

The company has been unparalleled at designing and exploiting such devices. In less than fifteen years Intel's sales moved from zero to over $1 billion. The company's products have consistently remained far in the vanguard. Even the most intense competition never seriously interfered with Intel's continuing successful scramble toward higher and higher product and sales performance.

However, by 1981—the year in which the events I will describe began—the Japanese semiconductor industry posed a serious threat to the fortunes, and even survival, of U.S. semiconductor firms such as Intel. Japanese companies had tremendous advantages over their American competitors. They enjoyed protected home markets, low-interest credit, government-provided or -backed investment capital, shared research, and development assistance. In effect, the Japanese firms didn't even have to worry about making a profit, certainly not in the short run.

The pattern had been for American firms to survive through advanced technical development. U.S. firms innovated, devising new products and manufacturing processes. When the new products reached the volume production level, the Japanese firms almost always were able to capitalize on their advantages to win the market battle for commodity sales. Japanese quality was high. Management effectively controlled costs to yield customer prices that American firms found hard to compete with profitably.

Typically, corporations in the United States made their major profits through a policy of skimming. They set high prices on products when they were introduced. The Japanese could not immediately produce the new semiconductors, because they had not yet copied the technology. Thus, the Japanese did not initially provide competition. In time, other manufacturers began producing the new product in mass quantities, bringing vastly lower market prices. At this point the original developer of the product often withdrew from the fray altogether and simply introduced a new, even more advanced product at a high, skimming price. Intel was especially likely to pursue this approach.

A good strategy, you might say. It simplified many management problems. And it produced an extremely good yield of profits to investment. The only problem was that it had begun no longer to work. The lead time for producing new semiconductors—and consequently the time available for the exploitation of skimming—was shrinking rapidly. Some products reaped only a few months of high-priced sales, not enough to reward the tremendous investment in development.

Here's a little episode related by Moira Johnston in *Esquire* (September, 1984, pp. 73–78) of what happened in those continuing skirmishes between American chip developers and Japanese chip manufacturers. Intel scored a major victory. It directly invaded the protected Japanese market and effectively side-stepped the tremendous threat of shortened development lead times.

In 1981, Intel was working on a chip to be used primarily for the relatively permanent internal storage of these instructions are for handy little operations that have to be performed over and over again in many different contexts. They are in effect clerical procedures such as might be found in an office manual. These procedures need to be kept quickly available for the processor at all times.

Up to that time, these essential routines typically had to be provided in a memory chip that was in effect custom-designed for every single application. Even fairly trivial design changes in a computer required a new chip. This burdened every new processor use with a tremendous up-front development overhead.

The new chip Intel was designing attacked this problem head-on. It was a general purpose program-storage device that could be adapted to work with any processor or application. The same product could be used for nearly any computer because the information the chip carried could be added (programmed) after the chip was manufactured. The information did not have to be part of the hardware. If the exact information required in the chip changed because of a design change in the computer of which it was part, new engineers did not have to develop new chips. Instead, the old information could be erased and new information programmed in. The chip was up to that time the most effective version of an EPROM, an eraseable programmable read only memory. Intel's name for the chip was the 2764 EPROM.

Intel called manager Bob Derby to its main headquarters to head the marketing effort for the 2764. Derby had been in Japan for nearly two years with the main goal of learning about Japanese business practices and culture. He had been so successful at assimilating Japanese culture that he was reluctant to leave. At the same time, he couldn't resist; Derby was being called home to battle. The opportunity to fight and win was compelling.

This is the situation he faced. Intel needed a new strategy to overcome the failing efficacy of their skimming policy. The 2764 was truly a hot chip and offered a good opportunity to initiate the new approach, whatever it was to be. At the same time, the chip had to be introduced right away to exploit available lead time. But Intel hadn't decided how to manufacture the chip or even whether it could be effectively manufactured. The production technology selected—the wafer stepper—had never been used in production runs outside the laboratory. The chip was scheduled for production at a brand-new plant that itself had never been shaken down to work out the bugs.

Derby proposed a leap of faith. Reverse nearly every aspect of the traditional product-introduction strategy, he proposed. Introduce the chip at a low price. Make sure that large quantities are immediately available at a quality level that will meet or beat what the Japanese can later deliver. Finally, introduce the product first to the Japanese buyers of chips. Invade the Japanese home market directly and aggressively.

Derby revealed his ideas to his boss, George Schneer, who explained them to the senior decision makers at the company.

Naturally, there was a good deal of consternation. The needed levels of production quality had never been achieved before. The needed yield of good chips to inspection rejects had never before been achieved early in a production run. The abandonment of price skimming seemed to be throwing away revenue. The government protection of the Japanese market had never been successfully challenged on the scale needed to make that part of the strategy work.

Instantly, it seemed, everyone in the company knew about the proposals and was talking about them. Naturally, there was hesitation. There was no immediate rallying around to accept such a radical departure. Derby had reason to feel even more alone in his ideas when he got specific about dollar amounts. He planned to guarantee a unit price of $16 within a year of product introduction. At that time, the normal Intel price would have been about $70.

However, when Derby approached Ed Gelbach, who had managerial control of the EPROM projects, about the price decision, Gelbach almost immediately gave his go-ahead. Others in the company at the same time were catching fire from the boldness of Derby's marketing offensive. Soon, practically the whole management team was behind the daring strategy.

A final component had to be dealt with. Could Intel manufacture the chip in sufficient quantities and with sufficient quality to profitably compete at the guaranteed $16 price? The response in effect was to look to human spirit to meet the challenge. The management team didn't assign experts to draw out new process charts and work flows; nor did they rely chiefly upon other engineering responses. The heart of the 2764 team's approach to the fabrication division consisted of an emotional call to battle. Derby and others felt that if the people in the production plant could be inspired with fervor for a head-to-head assault on the Japanese market, they would find a way to get the output and quality. Personal visits to manufacturing with repeated detailed discussions of the goals and general thrust of the marketing strategies followed.

It turned out the product could be manufactured. Yields and quality soon rose dramatically. Not long afterward, Fujitsu tested Intel's ability to deliver in quantity. They placed an order for 24,000 units with the condition that delivery had to be made within three weeks. A mad scramble at the manufacturing plant to collect every possible functioning chip succeeded in meeting the order. From that point on, success was to be assured.

Incredibly, electronic components follow many of the market laws of women's fashions. Fads, personal contacts and loyalties, and marketing verve seem to control sales of individual parts. The 2764, partly because of the energy and innovativeness of its introduction, became a best seller. The direct battle for the Japanese market was won, though temporarily. Intel maintained its dominance in a market that was critical to its survival.

MANAGING THE HIGH-TECH WORLD

What's going on in this story? Have we just seen a few hot-shots in action? Could their story have been repeated in the U.S. textile or auto components or agricultural processing industries? Maybe, but I don't think so.

Throughout this book we will look at some of the management attitudes and behaviors found at companies, such as Intel, that continually operate at the leading edge of technology. For the moment, note a few characteristics the Intel managers displayed that are not commonly found throughout American industrial concerns with sales in the billion-dollar range.

1. When the story of the 2764 opened, Bob Derby was assigned to Japan. He was there largely to absorb the Japanese atmosphere and to get an *intuitive* feeling for the ways Intel's major competitors do business. This sort of investment in giving managers an understanding of the human and cultural tendencies of competitors is a bold recognition of the true

importance of such factors in business success.

2. Innovation is a way of life. This is true not just in product design but also in the adoption of business policies for exploiting the products. "High technology" seems to have generalized in these managers' minds to refer to every business function. They apply the same creativity and intuitive leaps to marketing that they are used to in inventing new devices and processes.
3. Communication is more like that found in a really effective family than that of an authoritarian business or military organization. Managers communicate *in detail* and personally with other managers on every level. Knowledge of important initiatives seems spread magically to every edge of the organization. People are excited about and interested in what is going on.
4. It came naturally to these managers to issue a call to human pride and spirit. They did not attack their difficult manufacturing problem with management charts and disciplinary threats. Rather, they applied the belief that shared zeal for competitive success would release the natural abilities of the manufacturing group sufficiently to let them meet the tremendous challenge presented them.
5. These managers accept without apparent qualms a business environment that is continually threatening. It never seems to occur to them that there are other industries where the very existence of their company would never be in doubt.

6. Intel accepts risk: true risk that might better be called peril. The judgments involved in bringing out the 2764 were not simply about whether some alternative strategy might produce 2 percent greater revenues. Almost any action Intel took had considerable chance to produce actual permanent damage, not just a marginal loss of benefits.

One thing we are not saying is that these managerial attitudes and behaviors guarantee a company's success. The dominating element in any high-tech business is its riskiness. Even firms whose managers display all the characteristics we're discussing can fail or suffer serious setbacks. But what we are saying is that exceptional performance—without which no high-tech firm can succeed—is based on these managerial characteristics. And high-risk firms whose managers *don't* display these characteristics are almost certain to sink out of sight.

ACHIEVING EXCEPTIONAL MANAGERIAL PERFORMANCE Instilling Confidence, Pride, and Commitment in Others

6

The foundation of exceptional performance is *personal* involvement in the organization's goals. You need this involvement in your own work life and you need to engender it in your subordinates. Many managers even accept the responsibility of encouraging the involvement of peers and superiors.

This essential attitude toward work and toward the organization's goals and interests is built of three components:

1. confidence in one's ability to carry out assigned work adequately or excellently
2. a pride in the results of sincere efforts for the organization
3. a long-lasting commitment to the work content and to the organization

These characteristics—confidence, pride, and commitment—are clearly abstract bundles of skills, attitudes, feelings, personal relationships, technical and organizational knowledge, and personality traits. Some of these factors cannot be much affected by management actions. However, the key point identified by the most successful high-tech managers is this: *confidence, pride, and commitment can be built by consciously using sound management practices. You can augment your staff's personal involvement with the organization by using concrete, planned actions.*

Begin to explore the ramifications of these attitude factors by thinking of a counterexample. Recall the *worst* manager or employee you have encountered. This person may be someone you have worked with or a manager in a company or government agency that you have dealt with as a customer, service-seeker, or supplier.

Think carefully about what made this person a poor manager. Was it really that he or she didn't have technical abilities to perform the assigned work? Was it a lack of intelligence? Was it simply not knowing what to do?

Chances are, you will find that none of these factors is truly central to the poor performance of the employee you are thinking of. You will probably agree that the incompetent employee could have done the necessary work if only he or she had wanted to.

This is the central point. You not only have to see that your subordinates have the nuts-and-bolts skills of the job—that they are *able* to do it. Beyond that, and even more important, you must take steps to see that employees *want* to do the job. It is in this last area that the successful high-tech managers can contribute their most useful observations.

The surveyed managers identified four central areas in which your conscious efforts can best help build confidence, pride, and commitment:

- Provide experiences to learn from.
- Practice letting go.
- Use symbols and slogans.
- Create the excitement to achieve.

Each of these practices, of course, consists of a complex of management behaviors. Discussion of the practices in the final sections of this chapter will show how you can carry out a variety of activities to provide learning experiences, show that you have let go, exploit symbols and slogans, and create excitement for accomplishment. These practices will in turn increase the degree of confidence, pride, and commitment in the work groups you supervise.

Confidence

It is already widely accepted that motivation falls—often sharply—when an employee does not have the ability to perform assigned tasks well. This is a keystone of the expectancy theory of motivation. It also is just common sense. No normal person is likely to keep trying again and again to accomplish a task for which he or she lacks the needed skills. Even a highly skilled person will lose motivation if his or her supervisor wrongly perceives that the needed skills are unavailable.

Consider this example given by one of the survey managers whom we will call Charles:

> "I've done six projects for my boss over the past eight months. Each time he's told me that my work has been excellent. On the last two projects, my boss stopped by to congratulate me on what I did.

> "The project we are working on now is different. My boss keeps calling me up and telling me how to do my work. I can't understand what I could have done wrong. Doesn't he think I'm capable of doing my job anymore?"

Charles clearly thought that his boss had lost confidence in Charles's work. He interpreted the constant close contact as a sign that the boss thought Charles couldn't handle this latest assignment. Charles described his boss's feelings as an obstacle to good performance.

In fact, the boss hadn't lost confidence at all. On this particular project, the boss was getting a lot of heat from *his* superiors. His real concern was to keep close contact to satisfy his own managers.

No matter what the boss's reasons, the message he sent to Charles was clearly negative: "You're not trustworthy enough to handle this important project on your own." As a result, Charles lost confidence in his own work.

Confidence is a demonstrated faith in one's abilities, skills, and characteristics. This sort of confidence is personal confidence: "I know from past experience that I have what it takes to do this job well." The definition I will use includes a second aspect. There must also be organizational confidence: a demonstrated faith in the abilities, skills, and characteristics of subordinates to do good work in their positions. This might translate: "I know the people who work for me have what it takes to get their work done right and help me meet goals." Both kinds of confidence, personal and organizational, can be built.

Successful high-tech managers have found confidence especially critical in managing the frequently conflicting forces in high-tech industries and departments. Confidence promotes outstanding performance in any setting. It is absolutely essential in the environment faced by high-tech managers. The survey managers pointed out three main reasons for the necessity of confidence in this environment.

- High-tech managers and employees typically must rely greatly on their own skills and abilities. No organizational processes can be built to provide the sparks of genius that are so important to the success of high-tech undertakings. In traditional management, risk taking and inspiration are typically more centralized at the top of the organization. In typical high-tech companies, the risk taking and innovating are spread more uniformly through the organization.
- High-tech managers have an especially pressing need to rely on the knowledge and abilities of subordinates. These managers regularly work with subordinates whose technical skills and abilities far exceed their own. If the manager lacks personal confidence, this situation will be threatening. If the manager lacks organizational confidence, the subordinates will sense this and be hampered in their productivity. This situation arises far less often in traditional environments. There, the level of expertise tends to be more closely related to a manager's position in the hierarchy.

- High-tech organizations have usually applied decentralization to permit faster and more pertinent decision making. High-tech managers have been quicker to respond to the kind of judgment expressed in an article by Drs. Modesto and Hayes: "Anything that restricts the flow of ideas or undermines the trust, respect, and sense of commonality of purpose of individuals is a potential danger" (*Sloan Management Review*, Winter 1984). In order to lessen this danger, many companies have moved decision making down the line to the employee who has the greatest detailed knowledge of the issue. Decentralization clearly requires both personal and organizational confidence. An employee who is unsure of his or her abilities will be a poor decision maker. A boss who doubts the abilities of subordinates will be a poor candidate for passing important decisions down the line.

Pride

Contrast these two hypothetical situations. In both cases, George has worked hard, putting in unpaid overtime, to prepare a difficult report for an important meeting. George's boss, Susan, badly needs the report for the meeting tomorrow.

> "George, I really appreciate your getting this report to me on time. It's extremely important that I have it for the meeting tomorrow. Why don't I quickly look it over with you now before you leave? That way we can knock off

any questions right away." Susan silently reviews the report and continues: "It looks like this took a lot of work for you to pull together."

George answers, "It sure did, Susan; I worked over ten hours straight."

Susan says, "It shows; you have definitely exceeded what I expected. You not only gave me the conclusions I expected for each of these accounts, but you provided histories of each account with recent transactions. That's really useful. I want to ask you about this one section." After several minor points are discussed, Susan ends the meeting by saying, "You should be proud of this report, George! Thank you for doing such an excellent job."

Here's the second approach:

"Thanks, George, for bringing me this report on time. It's really important that I have it for tomorrow's meeting. Let me glance over it so I can ask any questions before you leave." Susan silently reviews the report and continues: "George, what about this table on page seven? The total column doesn't add up to 100 percent. I can't walk into that meeting with errors. George, will you make absolutely sure that all these errors are corrected before 8:30 tomorrow morning?" George is a little startled but answers, "Right, Susan." Susan then concludes the meeting by telling George he should be proud: "Except for the errors, the report looks pretty good. You did a fine job."

The difference in these two approaches is clearly one of emphasis. Yet the results of seemingly minor differences can be tremendous. In the first case, George left Susan's office feeling proud of his accomplishment. In the second case, even though Susan said George should feel proud, he didn't. He left Susan's office feeling disgusted. After his giving an all-out effort and working ten hours straight, the boss emphasizes a simple error and then implies that there may be many other such errors. George felt pride when he entered Susan's office. That pride had been destroyed by the time he left.

The survey managers gave many examples of the positive value of their subordinates' pride. They stressed that pride is essential to exceptional performance. They also repeatedly reported that managers' actions can be directly responsible for the level of pride—good or bad—that subordinates feel.

So pride is good and management actions can affect the degree of pride subordinates feel. What is pride? Why is it worth the effort to consciously build pride?

Pride is a sense of satisfaction received from a personal identification with something that is done well. It is brought about by being a committed member of a group that successfully struggles to meet goals and overcome obstacles. People may have pride in nearly anything under the right circumstances. Many are proud of accomplishments at work, of the people they associate with, of the life style they have built, and of specific possessions, such as automobiles and houses.

Pride is an emotion: it is something we feel rather than something we think. On the job, the anticipation of feeling pride is a powerful motivation. Pride in past accomplishments brings enthusiasm for facing new challenges. The expectation of feeling proud about a job well done is one of the most important factors in leading people to make personal sacrifices, display courage and toughness, and use the perseverance and endurance needed to achieve difficult goals.

This sort of pride was seen clearly when astronaut John Young first stepped from the space shuttle Columbia. Every mode of expression—his face, gestures, bodily posture, and words—showed the strong emotion he felt from the tremendous achievement of his team. On the Fourth of July, millions of Americans feel proud to be members of a large, successful group: citizens of a great nation. The U.S. Marine Corps speaks importantly to the pride of its members. The traditions of bravery, self-sacrifice, and dogged persistence are passed on to new groups of recruits.

In business management as in other fields, pride can be a great motivator. In a sense, it captures the emotional satisfaction of past successes and makes these good feelings available for facing new and difficult assignments. Pride in our organization leads to a further identification with the goals of the company. We come to feel that the organization is good if it can provide us with a continuing chain of positive feelings. Pride thus feeds on itself. Pride in the past gives us the strength to attack new challenges. These new challenges, when mastered, lead to yet more pride, in a continuing cycle.

Commitment

Here's a little-known story that demonstrates the level of commitment required to bring true innovation into the business environment.

In the early 1960s, a former radio and TV repairman decided to take on AT&T. Jack Goeken, the hero of this story, proposed going into direct competition with the long-lines service of the communications giant.

The move certainly seemed to be folly. Tremendous investment was needed for the communications plant. The management burden of installing a nationwide system from scratch seemed daunting to anyone but Goeken. Furthermore, the installation and operation of such a system was quite possibly illegal. AT&T was a protected monopoly. Direct competition had rarely been considered.

None of this stopped Goeken. He put together Microwave Communications, Incorporated (MCI), and together with William McGowan won the fight to compete.

What has Goeken done lately? For several years he has continued to struggle against the odds to bring to life Airfone. This service allows air passengers to make direct telephone calls to locations on the ground. Goeken, in partnership with Western Union, has put together a $20 million start-up operation.

In both of these undertakings, the key to success was persistence. Michael Bader, a member of the board of MCI, says, "Jack kept it going. He somehow kept himself alive, somehow getting materials together, getting people to cooperate, doing all that and somehow or other staying alive. Others wanted to give up, but Jack never seemed discouraged. He was always able to smile and joke about things." (*American Way*, October, 1984, p. 138.)

What Jack Goeken has is commitment. He has the drive to make plans work even when repeated obstacles arise. In the most successful high-tech companies and departments, this commitment will often be widespread. With the uncertainty of the high-tech environment and with the unexpected pitfalls of developing leading-edge applications and products, commitment is badly needed.

Commitment is a conscious pledge to oneself to pursue the path to an important goal notwithstanding the frustrations, difficulties, and obstacles in the way. Commitment is long-term. Commitment is the will to keep at it as long as necessary. Even when difficulties bring strong disappointment and apparent failure, the committed person carries on, using intelligence and endurance to forge achievements.

A person who is committed has a great willingness to perform work that really will contribute. He or she accepts group goals as being his or her personal goals.

According to the survey managers, the ability to remain committed through thick and thin grows from a deep personal involvement and identification with the goals and culture of the organization. The committed manager accepts the company's major goals as personal goals. Personal satisfaction—over a period of time—derives from organizational success.

One of the trainers for a Boston-based computer firm commented on one aspect of this commitment. People who work in high-tech environments, he said, have a strong interest in theory. The long-term dedication to the theoretical and basic scientific aspects of work allows them to contribute easily to current assignments. Each new assignment tends to be viewed as only a specific manifestation of the overall theory that truly holds the technicians' interest.

Commitment to the organization and to its tasks and goals must be augmented by another sort of commitment: commitment to the people in the organization. The successful high-tech managers repeatedly named this as a factor in success. Commitment to people is merely a description of mutual trust. The high-tech management style rests on risk taking, trying the untried, and making important decisions with sketchy, often inaccurate, information. A manager can do none of these unless he or she trusts others and is trusted by them.

A truck driver for one of the automobile companies described this kind of trust and showed what happens when the trust disappears. He told me about the great stress he and his coworkers were experiencing as they tried to adjust to a new boss. Sam, their former boss, had decided to retire about eight months earlier. Every person in the department had worked at least ten years for Sam; all of them would have done anything for him. Whenever he received any pressure from above about a particular shipment, the men would unite and get the job done. This deep sense of commitment was completely lacking with the new boss. The lack was showing in loss of enthusiasm and in less effectiveness on the job.

Circumstances almost exactly the same can be found in the armed forces. It is common to hear soldiers say that they will do anything for a certain sergeant. Such groups have often been noted for outstanding performance in battle.

During one of my interviews with Kay Stites, assistant district manager at Hughes Aircraft, we talked about commitment and how it relates to performance. "I have found, and continue to believe," he says, "that people want to feel good about themselves and want to do good work. When there is commitment, people always do more than you expect from them." It's this difference that leads to exceptional performance.

Are you committed? You should be interested in knowing. A good test is to ask yourself whether you would continue enthusiastically in your current field if the work didn't pay good money. If the answer is yes, chances are that you are highly committed to the work you do and get major rewards from it. You are probably able to define the many ways your work is important and worthwhile. Your work can serve as a path to betterment of yourself and of your organization, community, and society.

The meaning of commitment remains the same, regardless of the work setting. What one is committed to, though, may change as one moves from a traditional environment to a high-tech environment. For example, high-tech managers must be committed to a flexible approach to choosing methods for reaching goals. This flexibility is itself a goal in the high-tech environment, although it may never be recorded as a formal company policy. This dedication to flexibility is important because, as the work group approaches each new level of unknowns, the group may initially fail. But they must not fail in reaching the final goal. They must persevere.

Ron Rossi, an engineer at Ford Motor Company and a survey respondent, called this commitment to flexibility "a self-correcting guidance system." When an organization has high spirits, enthusiasm, and a personal dedication to goals, it will continually adapt when it finds itself moving in the wrong direction. As managers and employees reach stumbling blocks, the group doesn't lose its drive. It is commitment that leads the group to constantly change direction so that presently it will choose the right direction.

This flexibility in a high-tech environment contrasts with the relative rigidity of more traditional environments. In a traditional company or department, processes and activities for reaching goals are fairly well known. The drive for efficiency often demands that work methods and procedures are imposed on work groups by outside specialists. The group is denied the right to choose from among a number of alternative approaches for reaching company aims.

MANAGEMENT PRACTICE

It is unlikely to work when a manager merely says, "Be confident! Have pride! Show some commitment!" These, important as they are in understanding the foundations of exceptional performance, are no more than abstract terms. They must be made concrete before a manager can promote them. The successful high-tech managers surveyed agreed on four practices that combine to enhance employees' confidence, pride, and commitment:

- Provide experiences to learn from.
- Practice letting go.
- Use symbols and slogans.
- Create the excitement to achieve.

The following sections examine these practices individually. For each of the practices I have provided consistent information arranged in a highly structured format. This consists of a definition, a differentiation, an explanation of the practice's importance, a how-to description, and a brief note about beginning to apply the practice in your own management job.

MANAGEMENT PRACTICE 1: PROVIDE EXPERIENCES TO LEARN FROM

Definition

This practice is logically a part of the manager's staff-building responsibility. It includes four related activities:

1. Establish specific, achievable goals.
2. Provide work assignments or training experiences.
3. Give prompt, focused, positive feedback.
4. Renegotiate future goals.

These four activities form a cycle of continual improvement of skills and attitudes. When goals are renegotiated, new tasks or training and further provision of feedback will be needed.

The most common tools for providing learning experiences are

- task, project, and position rotation,
- coaching on the job and in preplanned meetings,
- formal training in an in-house facility, at outside workshops and seminars, and in college and university classrooms.

Differentiation

Companies—traditional as well as high-tech—must provide learning experiences for managers. There is, however, a subtle difference between the goals in a traditional setting and the goals in the high-tech environment. Typically the chief aim of learning experiences in traditional companies and departments is to impart specific technical or managerial skills and knowledge.

This goal obviously also exists in high-tech management teams. According to the successful high-tech managers surveyed, however, the priority goal in giving subordinates learning experiences is to increase the personal confidence of subordinates. Often these people have the technical and academic skills they need. Further experience and training then logically aim to improve the environment for exceptional performance. An important criterion, then, for this further work is increasing employees' confidence, giving them a greater willingness to use their skills in an innovative way that accepts responsibility.

Importance

Obviously, well-planned learning experiences will improve skills and technical competence. In companies striving for exceptional management performance, however, further benefits may well outweigh this immediate goal. There are several benefits you should keep consciously in mind when planning and carrying out learning experiences.

- △ Well-chosen learning experiences increase confidence in two ways. First, the subordinate feels more competent to exercise judgment and initiative. He or she gains improved skills and may appreciate your demonstrating that you care enough to go to some effort for his or her benefit. Second, by improving subordinates' skills you increase your own confidence in them. Providing structured, goal-oriented learning experiences is probably the single most important foundation for effective delegation. You will never delegate if you don't trust the people to whom you must delegate.
- △ Helping subordinates learn and gain confidence is a sure-fire way of communicating your desire for exceptional performance. The process of negotiating improvement goals is definitely a clear signal that you *expect* better and better performance. When giving feedback during or

after the learning activities, you have an excellent chance to reinforce the climate of exceptional performance.

△ Job and task rotation are particularly useful for sharpening subordinates' risk-taking intuitions. Often the degree of risk you or your company finds acceptable is very difficult to express in words or numbers. The only practical way to learn the acceptable upper and lower limits of risk is by observing others and through trial and error in assignments that are not likely to be catastrophic to the department or company.

How to Do It

Of all the management practices identified as crucial by the high-tech managers, providing learning experiences is probably the most "traditional." The very same approaches you may have learned in your traditional studies of the management process will work.

□ Negotiate goals; this can best be done as an adjunct to interviews conducted in the appraisal system or in management-by-objectives meetings.

□ Set a time-limited learning schedule; the activities should be written down with definite beginning and ending dates.

□ Follow up to verify performance in the activities; this means simply to be sure that subordinates actually take part in the planned activities.

□ Provide feedback that particularizes the learning to current organizational needs;

this can be done in both scheduled performance-review meetings and informal day-to-day contacts. The high-tech managers appear to prefer one-to-one meetings that are scheduled in advance as the need for review becomes apparent.

- Evaluate the extent to which the original learning goals have been met; this critical step is easily ignored. You and the subordinate must verify that progress toward exceptional performance has been made. Only in this way can future goals be set realistically.

The only real difference proposed by the successful high-tech managers is to keep in mind a subtle distinction of overall goals. The *primary* way in which learning experiences improve the performance of high-tech managers is in increasing feelings that the company respects their abilities and expects continual improvement of skills and performance, and that they are people with considerable skills and talents who should be confident enough to take risks, work toward goals, initiate, and accept responsibility.

Getting Started

Getting started is easy. Prepare for the next appraisal meeting with the conscious goal of establishing concrete objectives and schedules for learning activities. Chances are good that subordinates will cooperate enthusiastically. Don't let this opportunity for building confidence pass. If your company's appraisal system is not appropriate for this use, set up a series of meetings with subordinates with the express purpose of establishing learning goals and learning experience schedules.

Naturally, many chances for learning will not be anticipated in these meetings. If, however, you keep in mind the central aim of building confidence and a feeling of self-respect, you can turn informal work assignments and project transfers to good use.

MANAGEMENT PRACTICE 2: PRACTICE LETTING GO

Definition

Truly letting go is the high-tech equivalent of traditional delegation. It consists of giving subordinates the authority to make decisions and perform tasks of their own devising to meet negotiated performance goals. Letting go in the high-tech environment involves extension of the often narrow definition of delegation.

Differentiation

These are the main differences between traditional delegation and high-tech letting go:

- Letting go has the central objective of allowing subordinates to learn directly from the work they are doing. In traditional delegation, much of the evaluation of success is done by superiors. In high-tech letting go, the manager to whom a project has been delegated must accept the prime responsibililty of judging its quality and results.
- High-tech letting go strongly emphasizes sharing the glory for success. If someone in the stock room originates an idea for new and better packaging, that person, not the distribution manager several layers up in the hierarchy, gets the recognition and reward. The implicit as-

sumption is that people at any level, not just middle- and upper-level managers, can be successfully entrepreneurial. Let the spotlight shine on the person who actually did the work or contributed the idea.

- The superior manager relies more on a vision of accomplishment and performance than on quantitative short-term goals. Measurable goals are also important, of course. But the successful high-tech managers emphasized *trial and error*. Trial and error only works when there is some tolerance for making mistakes, even serious mistakes. In letting go, managers are accepting a more volatile work environment. Good innovations will come, but so will crises and problems. With the guiding vision of what the individual, department, and company should be like, the errors can be used in an active process of striving for goals. The traditional approach to delegation often stresses the use of a control system to head off errors before they happen. This sort of control is excellent for achieving efficiency but poor for encouraging innovativeness.

Importance

The importance of letting go can be expressed simply: it is essential to allowing managers and employees to gain the confidence and experience needed for exceptional performance. In the high-tech view, everyone should have the self-confidence and emotional security to accept risk and innovate. Neither of these factors can exist until subordinates have had the chance to "go it alone." Like learning to drive a car, management performance grows only from getting behind the wheel and learning to stay on the road. The best driving teachers take over only when catastrophe looms.

How to Do It

According to the survey managers, the usual rules of delegation provide the groundwork for letting go. These principally consist of the following:

- ☐ Distribute authority. Give subordinates the right to make decisions and commit resources to achieve the level of goals they have been assigned. Provide needed resources, even if it means undertaking complex negotiations with other departments.
- ☐ Establish complete tasks. Delegate the whole job even if this requires breaking up a large project into relatively small tasks that have a distinct beginning and end. Be sure to let subordinates feel that they have accomplished a definite goal after the work is finished.
- ☐ Define jobs. Make sure subordinates have continuing authority to carry out certain kinds of actions. Be sure they understand their responsibilities and authority.
- ☐ Communicate. Help others understand the vision that underlies individual work tasks. Let them understand the whole to which their part contributes. Provide feedback on the effect their results—good or bad—have on the rest of the organization.

- ☐ Set and use standards and controls. Everyone who is part of the delegation process must clearly understand what the ground rules are. Often this can be achieved only in a relatively formal manner: write down the goals of the delegated work along with the standards that will be used for judging the degree of success.

These are standard observations about how to be effective in delegation. They work well in a traditional or high-tech environment. However, the successful high-tech managers stressed two major points that are especially important for creating the conditions for exceptional performance.

- It is vital to pass *the whole job* to the subordinate, even if the delegating manager must take extra time to define complete subtasks with beginnings, endings, standards, and goals. Passing the whole job typically means that the person delegated to has direct control of *all* the resources needed to accomplish a definite goal.

  Consider, for example, one "whole job." A marketing manager needs to revamp the company's distribution system to industrial customers. The major goal is to reduce the time to ship orders to three days while reducing distribution overhead to no more than 2 percent of sales. This is a unified undertaking with a specific, easy-to-understand goal. *One* manager should have direct responsibility for meeting the goal. If the marketing manager chooses to delegate this program, he or she should give the chosen manager all of the authority and resources needed to accomplish it.

Needless to say, parts of this major program will need to be delegated. These delegated parts must themselves be whole jobs with defined goals. There may be, for example, need for a custom report on shipping rates by carrier with summaries of volume and distance discounts. The person who is assigned the report should have complete authority to allocate time and other resources to accomplish the goal.

Only when complete tasks are delegated is it really possible for subordinates to use their talents and initiative. Further, letting go of complete work units makes it almost certain that the glory of success will go to the person actually responsible for that success.

- Integrate the stages of letting go with the learning plan already developed: suit the level of delegation to the current skills of subordinates. Truly letting go on significant projects is in fact one good way to give subordinates the learning experiences they need. Putting subordinates in real control of significant work will increase their confidence in their abilities. Further, it is only through letting go that the delegating manager can build enough confidence in the abilities of subordinates to let them create, innovate, and control their own work.

  Clearly, letting go should dovetail with the planned learning experiences of subordinates. The success of assigned projects and goals should depend on skills that the subordinate possesses when the work is delegated. Learning will always take place, of course, but it is demoralizing to be assigned work that one has no idea how to begin.

Getting Started

As is true of so many management practices, the main thing a manager needs to do is *do it*! This is especially true of letting go. Three steps are often helpful in getting started.

- Analyze the past performances of your subordinates. This usually can be done as part of the formal appraisal system. Try to come to an accurate judgment of the subordinates' past successes and failures. Write down notes to help you classify current levels of expertise and management skill.
- Take the time to define a single, unified, whole job. Think especially of goals and standards. Make these as clear as possible. Think of examples of good and bad results related to the goals, and use these examples to communicate the delegation to the subordinate.
- Carefully analyze the resources needed to accomplish the delegated work. What sort of personnel will be demanded? Is a budget necessary? What outside departments will help or hinder the work? What authority for making decisions and allocating resources is needed? When you determine the resources necessary, make sure that the

person delegated to has *direct* control over the resources.

MANAGEMENT PRACTICE 3: USE SYMBOLS AND SLOGANS

Definition

A symbol is an image or artifact that represents something other than itself through association, resemblance, or convention. A slogan is a saying that capsulizes an important value or goal. "Better Living through Chemistry," for example, was an outstanding verbal expression of the overall mission of one of our corporate giants. When employees accept such a slogan, their constellation of motivating factors broadens greatly. They can become committed to something beyond the numbers.

A certain department may have a goal, for example, of reducing manufacturing waste by 7 percent. This is a legitimate, specific goal. It is not inspiring, however. A slogan such as "Better Living through Chemistry" lets employees identify with a broader goal: improving conditions in the overall society. Such a goal is inspiring; it creates pride in doing important work; it provides personal justification for the extreme commitment to goals required in exceptional performance.

A symbol is a concrete expression or object that summarizes and stands for a group of values and achievements. The national flag, for example, can bring tears to the eyes of the patriotic because it recalls all of the struggles and accomplishments of the American people. Conquering the wilderness, bitter wars, surviving the Great Depression, and the ambiguous strivings of recent history are all represented by a piece of cloth of certain colors and shapes.

In many large corporations, something as simple

as the company logo used on stationery and in TV and print advertising comes to stand for the success and prominence of the firm. Employees and managers are glad to identify with this prominence. They come to enjoy seeing the logo and to identify with the company values it represents.

Differentiation

All human undertakings rely to some extent on slogans and symbols. Certainly companies in the traditional environment do. But there is a special need in high-tech companies and departments, for two main reasons:

- Many high-tech departments and companies are simply too recently formed to have developed the folklore and rich history with which employees identify. A substitute is needed.
- It is especially important for high-tech managers to have a strong commitment to and pride in the organization. When exceptional performance is demanded—and it is almost always demanded in the high-tech environment—calls to human emotional involvement are essential.

Importance

Let's start at the beginning. An important basis of our corporation-based economy is the proposition that groups of people can perform more efficiently than single individuals working alone. Work groups are the very heart of modern commerce. The groups may be small, as in a machining department with only six or eight operators. But each of these small groups is a member of larger supergroups extending in size up to membership in the international business community.

Human groups undergo considerable stress caused by outside threats and challenges and by internal strains arising from personality conflicts and competition. Many elements are needed to keep a group coherent in the face of these difficulties. One essential factor is a set of slogans and symbols that remind individuals of the importance and effectiveness of the group as a whole. When disagreements arise, symbols can effectively recall for individuals their past dedication to the goals of the group.

Further, symbols remind the group as a whole of their common goals. Slogans and symbols provide a human context for work goals. They point to entities—social good, cooperative satisfaction, overcoming obstacles, mutual striving, helping others—that are above our immediate personal interests and needs.

How to Do It

To use symbols and slogans successfully, there are two basic rules to follow:

1. *Communicate the folklore and oral history of the company.* If people have pride in their organization, they will remember those aspects of the company's history that reinforce that pride. It's common for people to remember and talk about those past events that exemplify heroism, triumph, achievement, and courage. Texas Instruments employees may take pride in the company's courage in flexibly changing direction and shifting product lines to streamline competitive power. IBM employees may be proud of their company's tremendous success in introducing its personal computer. This product attacked a market that was entirely new to IBM; yet the company's manufacturing, advertising, distribution, and support decisions gave them dominance in the market almost immediately.

Many companies have individual heroes, like the men who risk their lives trying to save coworkers trapped in a coal mine. Chrysler Corporation has a hero at the top in Lee Iacocca. He turned the company away from almost certain disaster. IBM has a personal hero. I have observed that people within IBM know Tom Watson's career history and talk about him as if he were a personal friend.

2. *Sponsor company rituals.* Rituals are another type of event people like to talk about and take part in. Golf outings, bowling leagues, softball teams, award nights, and community open houses for husbands and wives all help increase group cohesiveness. One start-up high-tech company, after its second year in business, began having weekly barbecues in the summer. Each Wednesday at lunch time the company would supply its over 600 employees with hot dogs and hamburgers. Managers prepared the grills and cooked the food for everyone.

Other companies sponsor a parade or company fair each year. Others yet have a standing practice of recognizing the birthdays of employees. They may hold birthday parties or present token gifts to mark the day. These examples support the second rule.

Getting Started

Three types of effort will help to get started in the effective use of symbols and slogans.

○ Review your current orientation process. Chances are that your new employee orientation could do a better job of communicating the informal history of the company and of making employees aware of important rituals they may take part in.

- Let employees at all levels know what is going on with the company. Communicate successes and failures. Recognize courageous or exceptional work. Make sure that everyone has a good familiarity with current company history.
- Establish group rituals. These should, of course, be matched to the interests of employees. One high-tech company (and a number of universities as well) sponsors a yearly contest for radio-controlled model airplanes. Such an activity is ideally suited to the predispositions of the many engineers working for the firm.

MANAGEMENT PRACTICE 4: CREATE THE EXCITEMENT TO ACHIEVE

Definition

Excitement to achieve is a human feeling that arises from signs that employees belong, are important to reaching goals, are appreciated, and are fully knowledgeable about what the organization is striving to do. In the ideal organization, according to the survey managers, results are what count, but it is the enthusiastic involvement in day-to-day work that gets results. Work and membership in the organization are important sources of excitement and stimulation for exceptional performers.

In simple terms, a person with the excitement to achieve typically feels, "I belong here; I'm an important part of this group, and we're really going to make our goals." A person who lacks this excitement will usually feel, "If I left here tomorrow, nobody would realize I was gone."

Differentiation

In the genuine high-tech environment, hardly any work assignment will be truly routine. Even the accounting and record-keeping function is likely to present challenges, because the many years needed for working out bugs are lacking. Nearly every employee is faced with constant new problems, shifting market and product factors, and other characteristics of the high-tech world.

This contrasts sharply with the situation in a typical traditional industry. In the traditional company, *most* work is routine. It can be efficiently performed by following procedures that have been refined over the years. The need for a highly stimulated management team is significantly less in long-established companies for whom the overriding goal is efficiency of production and distribution. But although most managers in traditional firms may not try to stimulate excitement to achieve, that doesn't mean that employee performance could not be enhanced if they did use this approach.

Importance

In a real sense, the excitement to achieve is the culmination of all the management practices. If well-trained, innovative employees in a congenial work group have this excitement, their chance of realizing exceptional performance is good. It has been repeatedly demonstrated that this drive, dedication, and entrepreneurial outlook can overcome many internal and external obstacles.

How to Do It

If excitement to achieve is the culmination of all the background factors for exceptional performance, it stands to reason that using the first three management practices will partly produce the desired result. A staff that has had proper learning experiences, whose manager has let go, and whose company environment contains proper slogans and symbols will already have much of the required excitement. Some specific actions to create the proper attitudes will reinforce and expand upon earlier techniques.

A manager should encourage and provide opportunities for employees to describe the results of their work to others in the company. This can take the form of an article in a trade magazine or in the company newsletter. Subordinates can also give presentations for other professionals in the field or for the public.

Slogans aimed specifically at creating the excitement to achieve will often succeed. A good example is the slogan used by the people who work with Mike Galle at AT&T in New Jersey. Mike told me in an interview that he and his people identify themselves as a group that does its work "Quick and Slick." People in the department remind each other of the time urgency of work, of its "Quickness." At the same time they keep sight of the need for exceptional ingenuity and quality, or "Slickness."

Posters serve to remind employees of important symbols and slogans. Warren Hoyt, Executive Director of the Michigan Press Association, was quick to point out during our interview that he had noticed on a recent foreign trade trip to Mainland China a tremendous number of posters displayed throughout workplaces. The posters all symbolized the value and quality of the work to be done.

Other managers create excitement by giving praise. It is well documented that people need praise. Well-chosen, genuine praise contributes to confidence, pride, and commitment. Praise should be worded to inform employees that they have made an indispensable contribution to causing something important to happen. In effect, their managers should tell them that the goals would not have been achieved without the employees' contributions.

Praise can backfire if it is not genuine. This happened at a research lab where I provided training programs. This division had a very profitable year, and it was rumored that their success helped materially to make up for losses in other divisions. The laboratory responded by sending a formal, standard memo to all employees. This seemingly uncaring gesture was not appreciated. One scientist told me, "It would have been better if they had done nothing than to do what they did. It would have helped if they had at least put our names on it."

Here are some tips for giving praise as formulated by the successful high-tech managers surveyed:

- ☐ Aim the praise specifically to the individuals and groups responsible for the success to which you are referring.
- ☐ Aim the praise to the specific actions and efforts that brought success. This makes the praise more concrete and credible and increases its value as a learning tool.
- ☐ Don't be afraid to make the praise public. It's criticism that must be kept confidential, not praise.
- ☐ Use formal, public reward programs, such as employee-of-the-month registers, salesperson-of-the-year banquets, and so forth.

Finally, remember the little things. A manager at Gould, Inc., Computer Systems Division pointed out to me, after turning in his completed survey form, that seemingly minor matters in the organization can affect a person's choice of career path. They can be like the subtle status symbols accompanying the management role that are often absent in the technical role. For instance, managers frequently are issued business cards, and technical people are not. "I realize that is a small difference," he said, "but when you look at the number of technicians that elect to move into management within the organization, it makes you stop and wonder."

Another person I interviewed commented that he had turned down a job offer from a Fortune 500 firm because the company gave the impression of being impersonal. Wherever he went in their offices, he saw only room numbers on the doors—no name, title, or anything else. The impression given was that the employees were nameless work units that could be plugged in or scrapped whenever convenient.

Getting Started

The first place to look when planning to encourage the excitement to achieve is within your own feelings. Do you have this excitement? Is your enthusiasm thoroughgoing and genuine? If not, you must consider what is lacking in the management teams above you in the organization. You may be able to use your persuasive and management skills to make changes that will cause you to feel more integral to organization goal-seeking. If not, you may have to accept the possibility that it's going to be hard to get others to feel excited when you yourself do not.

Assuming that you do have the excitement to achieve, how do you get started in communicating it to others? Review your efforts to provide learning experiences. Check your ability to truly let go and let recognition shine on the people who most directly deserve it. Review the set of slogans and symbols you use; if they're not working, change them.

Finally, review the current situation in the area you manage. What recent results have been produced for which the producers got inadequate praise? Chances are good that you will find some. Note on your calendar your plans to see that proper, genuine praise is offered.

ACHIEVING EXCEPTIONAL MANAGERIAL PERFORMANCE 7

Engaging Others in the "Vision" Through Expectations

When General Motors spent $2.6 billion to acquire the Dallas-based EDS, a successful data-processing firm, some analysts, including a number of GM managers, feared that the giant auto maker would suffocate the entrepreneurial spirit of the much smaller, high-tech EDS.

Such fears were unfounded. Instead, some insiders at GM feel that the question has become not how much GM will change EDS but how much EDS will reshape GM.

In the first place, nearly ten thousand data-processing employees at GM offices and plants became EDS employees. They lost their identity as GM employees; their fortunes were to be tied strictly to EDS, which is taking over all GM data processing. As EDS employees, many of these transferred white-collar workers were in for culture shock.

The new EDS employees have a different benefit package. EDS has a dress and conduct code. The company expected no facial hair, for example, greater than a well-trimmed mustache. Automatic pay raises were out; EDS paid bonuses and salary increases strictly to top performers.

Employees who had been with GM for decades found these changes hard to swallow. Some quit. Others sued GM over changes in pension benefits. The move created quite a stir inside GM.

Why was there a problem? Quite simply, the employees' and managers' expectations no longer jibed. Employees expected one sort of treatment and received another.

Society in general, and notably the society of the work world, is filled with expectations. Improperly managed, these expectations are the greatest obstacle to exceptional performance. Properly managed, expectations can produce a wave of strong, personal motivation that carries the organization toward its goals.

An *expectation* is the mental anticipation that someone will behave in a certain way. People express expectations with statements such as, "I expected you to do ________ ." The blank can be filled in with myriad professional and personal actions: finish the report on time; get the salary committee to approve a pay raise; get John's productivity up to standard; inform top management of how well our work group is progressing; and so forth.

The key to gaining the most motivation from expectations is to achieve congruency. *Congruency* exists when the expectations of two or more people agree on major issues. If a boss and subordinate agree on goals, tasks, rewards, methods of interacting, resources available, and the hundreds of other factors that contribute to effective work, they are said to have congruency of expectations.

Congruency sometimes is obtained by accident. Two people may have personalities, experiences, and training so similar that their expectations will basically agree. Don't count on this, though, because it is rare. The exceptional performing managers take conscious actions to achieve congruency of expectations. This is one of the most important tasks of the pure management role.

Sources of Expectations

The likelihood of obtaining congruent expectations by chance is made small by the existence of a number of sources and types of expectations. All of the expectations interact, creating an extremely complex situation.

Let's look at a scheme for classifying and describing expectations. Figure 1 shows six categories that operate in nearly every management situation. The following sections describe the six cells of the expectation matrix in more detail.

Expectations may be spoken or unspoken. When they are spoken, they are openly communicated, understood, and mutually accepted. Expectations that are unspoken may never be directly communicated in any form. These expectations typically relate to the personality and history of both the organization and its employees. But the unspoken nature of these expectations does not make them less important. In fact, leaving a large group of expectations unspoken appears to increase their importance and impact. The impact is rarely a favorable one, however.

Figure 1. The classification of expectations

	Spoken	Unspoken
Organization	A	B
Manager	C	D
Employee	E	F

A. Behaviors explicitly requested by the organization: production goals, budgets, policies, procedures.

B. Behaviors expected by the organization but not consciously requested: characteristic style of clothing, respect for privacy, expectations about when to arrive at work, etc.

C. Behaviors and responses that managers say they expect from the organization and from subordinates: sufficient resources, adequate salary, willing communication, etc.

D. Behaviors and responses managers expect from the organization and from subordinates but do not explicitly ask for: methods of working together, respect, accurate information, etc.

E. Behaviors a subordinate requests from the manager: how often to check with boss, who to go to if resources are inadequate, etc.

F. Behaviors a subordinate expects but does not directly ask for: respect for personal interests, consideration for promotions, protection from higher-ups and peers, etc.

Spoken Organization Expectations The explicitly stated expectations of the organization usually relate directly to what tasks will be undertaken and how the work will be done. Written or otherwise formalized policies, procedures, and work rules are all spoken organization expectations. These expectations are usually discussed up front when a manager is hired. Further specification and elaboration will continue throughout the orientation process and later in daily work.

The company and the employee discuss such matters as job qualifications, responsibilities, authority, work schedules, salaries, benefit packages, attendance requirements, check-off and approval mechanisms, and other factors related to the formal structure surrounding the job. All the information in the employee handbook and policies manual falls into this category. Any statement that expresses what the organization wants the employee to do and what the organization will do in return is an example of a spoken organization expectation.

Unspoken Organization Expectations The expectations in this category result from the cultural environment of the organization. Unspoken organization expectations encompass such elements as rituals, traditions, mores, and the unwritten rules of the organization.

One of the survey managers whom I shall call Bill told me of an experience he had that is a good example of unspoken organization expectations. Bill joined one of Detroit's largest manufacturing companies eighteen years ago. He recalls having lunch for the first time with several of the company executives. Bill was eager to advance within the company. At the same time, he confesses,

> "I was somewhat naïve. I will never forget when one of the executives next to me asked if I played golf. Unaware of the significance of the question, I blurted out 'No.' Apparently it was louder than I realized because an executive at another table shouted out, 'He will!'

Sure enough, the next day I was out on the golf course learning to play the game. I decided that if I needed to learn how to play golf in order to advance in the company, I was going to learn how."

Another example shows how far these unwritten rules can go. In one manufacturing company in the springtime no one is expected to wear white shoes until the company president does.

Many of these unspoken organization rules are rather trivial. They mainly serve the purpose of reinforcing employees' membership in the corporate "club." In many cases, however, they also encompass a far deeper meaning. For example, most companies have an unwritten rule about when one can and cannot go over the immediate boss's head to make a request of top management. A manager who is ignorant of the conventions can seriously damage his or her career hopes by violating the rule. In the tightly knit social group that exists in most corporations, such trespasses are not soon forgotten.

Spoken Manager Expectations A manager has spoken expectations aimed in two directions. In the first place, he or she will have a group of wants and needs that the organization is requested to satisfy. These include personal needs, such as respect and confidence. They also include the resources to do the job: budget, resolution of problems, expert assistance, moral support, and protection from interference by others.

Managers also have a second set of spoken expectations. These revolve around what the manager expects his people, his subordinates, to do. The expectations relate to fairly concrete assignments of tasks to be performed. The goals, assigned procedures, allocated resources, assigned staff, and so forth make up an important part of the manager's expectations. Often the manager's spoken expectations will also include factors such as how frequently to report on progress, what to do if a serious problem is encountered, and how and when communications are to be made from the subordinate to the boss.

An example given by one of those surveyed is the manager of the human resources department at a large hospital in Ohio. Jane, who reports to this manager, made this observation:

> "The boss expected each of us to pitch in and see a project through whenever we got caught in a bind. In return, he would not adhere closely to the hospital's rules for being on time. The hospital had a policy that everyone had to use time cards. Every department was supposed to penalize every employee who came to work late or took more than the allocated thirty-minute period for lunch. Our manager was more lenient. If we were five miutes late in coming back from lunch, he would not penalize us. On the other hand, when we were working on a tight deadline, he would expect us to stay and work on the project until we finished."

Unspoken Manager Expectations As with spoken expectations, managers have unspoken expectations of both the organization and subordinates. Many managers, for example, have an unspoken belief that their boss should and will help them advance in the company. Managers often have the unspoken belief that if they get seriously ill, the rules on sick leave should be bent. Others may feel that they may legitimately take it easy at work for a few days after a major project is successfully completed. Many managers expect the organization to give them the chance to work in certain areas of interest even if these areas do not currently agree with the major thrust of the department's work. Others feel strongly that the organization should make them part of the decision-making process even if their job title would not ordinarily permit this.

Managers have unspoken expectations relating to their people. Most managers feel that subordinates should always make the boss the first to know of any trouble; subordinates are expected to keep the wraps on a problem until the manager has had a chance to develop a response. Managers often expect personal loyalty: if it's good for the boss, do it; if not, don't.

There is a third area of unspoken expectations managers have—expectations of their own performance. These include the number of hours managers are willing to work, the extent to which they will protect the boss, the sort of rewards they expect for their work, the extent of supervision they feel comfortable with, and similar job factors. Even a manager who has considerable self-knowledge and aggressiveness will nearly always experience such factors without openly stating them.

Spoken Employee Expectations These expectations are the ones employees ask about. For example, especially during their early months with the company, employees ask many questions to pin down exactly the amount of authority they have for making decisions and allocating resources. They want to know how often they must check with the boss before going ahead. They ask how much direct responsibility they should take for solving problems. They want to know how much latitude they have for disciplining others and making their decisions stick.

Unspoken Employee Expectations These are the concerns employees do not ask about. They are almost exactly similar to the unspoken manager expectations. Will I get the rewards I want? Will this boss realize that everyone makes mistakes? How much can I move funds between budget line items? Do I have to work myself to death or can I spend some time with my family? What does it take to get a promotion around here? What really affects the amount and timing of pay raises?

Establishing Congruency of Expectations

Here is a true welter of expectations. Is it any wonder that expectations rarely match up unless someone makes a conscious effort to see that they are in agreement? It is the process of achieving agreement—congruency—that the exceptional performing managers most often addressed in their survey responses. Managers have talked for years about how to harmonize the goals of the employee with the goals of the company. The three practices discussed below are the exceptional performers' answer to how to do this.

MANAGEMENT PRACTICE 5: IDENTIFY THE EXPECTATIONS OPERATING IN A SITUATION

Definition

If you wish to manage and build expectations you must first decide what important expectations are operating from the organization. The importance of particular expectations depends, of course, on where the company is heading and on what true product or results you are seeking to produce.

The most important task in identifying organizational expectations is to determine:

1. The true vision pursued by the organization.
2. The methods of working together that are genuinely supported by top management.

Differentiation

Traditional industries and companies have a more homogeneous make-up. Long-established traditions relate to many areas that prompt important expectations. Even when organizational expectations are not communicated through explicit statements, they are often so clearly communicated in other ways that incongruity of expectations is less of a problem. Employees who cannot harmonize their goals with the structured goals of the organization are relatively few. This is because, at the very beginning of the recruitment and selection process, potential employees are constantly indoctrinated in company lore. The few who cannot or will not accept the organization's and manager's expectations can simply be discharged. The loss will be relatively slight.

Identifying expectations turned out to be an extremely valuable experience for a small group of managers who worked for a large, nationwide, telecommunications company. The managers were participating in a training program I was conducting. One of the activities involved having the individual managers identify the organizational and managerial expectations operating within the company. When the managers completed the exercise they were surprised to discover that the expectations they had identified were not congruent with those of top management's. The activity gave the managers the opportunity to start redirecting their energies in a positive, more meaningful direction.

Typically, in the high-tech environment, there are no long-standing traditions that have been built. There can truly be a chaos of expectations because every individual will have a slightly different spin on what is acceptable and what is unacceptable. This same lack of standing traditions is typical of the advertising profession. Joe Bride, Jr., vice president and general manager of Lawler Ballard Advertising Public Relations Institute, during an interview explained that "working in ad agencies is similar to working in high-tech because we are presented with challenging projects with no known past results." Harold Shoup, president of Carr Liggett, a Cleveland-based advertising agency, concurs with this. "Furthermore," he said, "we operate with no inventory. We are always creating something new. Thus, it becomes imperative for us to effectively utilize the resources within our agency."

Employees and technicians who have trouble harmonizing their own goals with those of the organization cannot usually be discharged unless the problem becomes very serious. Many high-tech companies and departments cannot afford to lose technical expertise and specialized skills without damaging output. It thus becomes more important in the high-tech environment to find a way to make expectations congruent.

Survey manager Walter Work, Jr., of Federal-Mogul mentions another difference. A traditional environment, he says, is thought of as a system in which everything is justified economically. In high-tech environments emphasis is placed upon viewing the organization as a total system. "The organization," he said, "stresses optimizing all its resources, which include the social, political, technical set of systems in addition to the economic system."

Importance

One goal when managing expectations is to make employee expectations congruent with organizational expectations. You cannot do this until you get a clear picture in your mind or in writing of the types of expectations operating.

How to Do It

The long-term goal of managing expectations is to consciously establish expectations that can be discussed and enthusiastically accepted by you and the greater organization, and you and the people who report to you.

The beginning stage of this essential process is analyzing current expectations and desired future expectations that exactly support the organization's vision of its product. After this process, you should know the kind of expectations you wish to establish to support the overall vision.

Your effectiveness in building the right expectations depends on having a clear view of what the organization needs and wants. You will need the support and clarification of upper-level management in two areas.

1. Clarify and verify the overall vision of corporate success Every manager has been repeatedly advised, "Know what your organization's goals are." This, of course, is good advice. To achieve exceptional management in yourself and in your people, you need to know the overall vision.

Vision is different from goals, however. Goals only encompass rather narrow accomplishments, usually expressed in concrete numerical terms. The vision transcends these. Goals are formulated in order for the company to achieve its vision.

Robert E. Kelley, in his most recent book, *The Gold-Collar Worker*, further illuminates the importance of vision: "Whether a vision is a matter of creation or discovery, it inspires an entire organization and infuses it with a welcome sense of direction. Many experts believe that followers within an organization that clings tenaciously to its vision are more contented and more productive, and that the organization itself meets with greater success."

U.S.A. Today, for example, has the overall vision of being the second-most-widely-read newspaper in the country. This vision expresses a strong immediate statement of the newspaper's desired placement in the market. Contrast with this a goal: increase circulation by 10 percent this year. This is a target that has meaning only in relation to the overall vision.

At Meijer Thrifty Acres, a successful chain of over fifty food stores, the stated vision is "Our product is our customer." Brian Michmerhuizen, a manager surveyed in the study, and the training and employment manager at Meijer, says, "When we communicate the vision to our people within the organization, everyone can identify with it and feel that they can be a part of it."

2. Learn directly from top management how they want people within the organization to work together Top management sets the tone for the whole organization; this is as it should be, since top management is most directly responsible for accomplishing the company's vision.

Walt Work at Federal-Mogul contends that a manager's effectiveness in creating congruency within an organization depends to a great extent on lower managers being able to learn and put into practice the view of the organization that has been formed at the top.

Specifically, do the leaders want to encourage diversity of thinking and openness of expression? Do they want their people to engage in important problem solving? Would they prefer other managers just to provide information and leave the decision making to top management?

The expectations you will establish should be controlled by the true expectations of top management. For example, a middle manager of a Michigan-based company wanted me to train supervisors and their employees. The goal of the training was to open up communications and make both the supervisors and employees more participative in the decision-making process.

After I had discussed the program with the manager in more detail, it became apparent that the participative approach was inconsistent with actual corporate practice. The current management structure was strongly top-down. Decisions were made by the people at the top and then filtered down to others in the organization for implementation. If I had offered the program, before long the trained employees would realize that the structure of their company actually discouraged participation and decision making. The training would have had no value because it was impossible to apply its results.

Getting Started

For practice, pick any task or area of work in your operating area. List some of the spoken and unspoken expectations of the organization, yourself, and your subordinates that apply to that situation.

Further, describe in writing the two factors that determine the expectations you will try to establish:

- the true vision pursued by the organization,
- the methods of working together that are genuinely supported by top management.

To put these realizations in practice, you will need to create events (discussed in chapter 8) such as meetings or the establishment of motivating situations.

Management Practice 6: Make Expectations Explicit

Definition

The discussion and mutual acceptance of all the expectations operating in a situation create a firm agreement among the people working in the situation.

The manager can express his or her ideas and requirements related to the work of the employee. The manager also can greatly clarify the relationship between the operating group and the whole organization. The manager's employees can express their expectations about how they will work and about what they will get in return for the work.

The immediate goal of making expectations explicit is to make formerly unspoken goals spoken. Congruency can never be achieved as long as expectations remain unspoken.

Joe Bride, Jr., of Lawler Ballard, said in our interview that he works to establish congruency by establishing clear expectations before "we move from the strategy stage of a project to the execution stage."

> "Rather than starting with the objectives of the project we begin with gaining an understanding and a clarification of everyone's expectations. We do it at an open session with all the players present, and together we communicate what each person expects from the project. The advantages of taking this approach are many. One in particular that stands out is that by having everyone communicate what they expect, the person who is off target will be able to easily identify it for himself. Also, by focusing on defining the direction, false expectations are noticeably reduced."

The process of bringing unspoken expectations out into the open can be quite difficult, even grueling. The problem is that even the person holding the unspoken expectations may be essentially unaware of them. The only way to overcome this difficulty is through frequent and copious communication.

After unspoken expectations are brought into the

open, a process of negotiation takes place. Since new expectations will continually be aroused by the ongoing work of the organization, this negotiation will be more or less continual through the years.

An organization that is held together by a common set of known expectations may be called *in phase*. Creating an in-phase organization is the goal of the process of making expectations explicit.

Differentiation

In traditional companies, every employee is likely to have many models from whom to learn expectancies that are consistent with the goals of the operating unit and the vision of the organization as a whole. This means that in traditional companies it is not as essential to bring unspoken expectations into the open.

This multiplicity of models is less likely to be available in a strictly high-tech company or department. Individuals are encouraged to act entrepreneurially. Thus, there is a tendency for each employee to negotiate his or her own conditions of work and compensation and benefit packages.

Under these circumstances, someone has to make a conscious effort to harmonize personal goals and expectations with those of the manager and organization. In a high-tech company or department, this harmony is essential. Making expectations explicit is thus a more essential step in producing exceptional performance in high-tech environments.

Importance

Without congruency of expectations, long-term exceptional performance is impossible. Unspoken, implicit expectations form a barrier against nearly every feature and condition that produces exceptional performance. Communications will be oblique and incomplete. Confidence, pride, and commitment will be undermined. Conflicts will be frequent because all parties with unspoken expectations will have, in effect, a hidden agenda in every work situation. Loyalty and enthusiasm will suffer because managers and employees find that they are not consistently getting from work what they want.

How to Do It

Obviously, frank and open discussion of expectations must be at the heart of efforts to make expectations explicit. However, some intermediate steps smooth the process of bringing desires and anticipations into the open.

1. Create enthusiasm in yourself The first need is to create and maintain your own enthusiasm about your point of view. Your enthusiasm will be infectious, and others will be more enthusiastic and more highly motivated to find a way of harmonizing expectations. The situation is similar to a frequently stated maxim of selling: "It is hard to convince others if you are not first convinced."

Part of the need to create your own enthusiasm derives from the expectations of the organization that you do not wholeheartedly accept. For example, your company may establish a policy that you disagree with but are still expected to implement. You must accept ownership of and the responsibility to carry out the policy. You must support the policy when communicating it to others. A disbelieving boss will create disbelieving employees. Your people will not perform exceptionally when they get mixed communications.

Ralph Nichols is president of the Ralph Nichols Corporation, which sponsors Dale Carnegie courses. In his interview, he indicated that he uses a process of "envisioning." He considers a project at hand and pictures it in his mind to get a feeling for what will happen when the project is completed. Ralph says that while mentally viewing the completed project, he begins to feel that bubbly feeling of enthusiasm. From years of experience, he has learned to recognize that his excitement is beginning all over again and he will be ready to tackle the project.

2. Engender enthusiasm in others Expose your own enthusiasm so others will have the chance to "catch" it. Make plenty of contacts and express your enthusiasm about current projects. Smile, modulate your voice, and listen carefully to what others say. Remember, suggestions others make are indications of their expectations. They wouldn't comment on their work if they didn't expect you to take it seriously.

Asking the right kinds of questions can convey enthusiasm and bring out expectations. An example is the question, "If we were to forget about the pay, what about the work in this project is exciting for you and brings a sense of satisfaction?"

The GM Orion, Michigan, plant is a good example of a work force with a proper enthusiasm for the company's vision and goals. The plant builds the midsized Cadillac. The human resources people there said during an interview that everyone around the plant talks about "our new, beautiful plant and the wonderful cars we build." In the same breath, they also talk about the people who work there and how important it is to treat others with dignity and respect. It is apparent that receiving and giving such treatment satisfies the spoken and unspoken expectations of employees.

3. Reduce uncertainty Clearly, the reason that expectations are often incongruent is simply that managers and employees do not know the expectations of the boss or of the organization. In some cases great uncertainty exists. Employees legitimately ask, "Why am I doing this work? What's the good of it?"

The answer to this problem is distinctly to tie individual work goals to the vision of the overall organization. Make sure that everyone knows how his or her job contributes to the joint efforts of the whole organization. The manager should communicate that an individual job is essential to the company's achievements. If it is impossible to do this, probably the job should never have been created in the first place. Every job should be genuinely needed.

Jim, a top salesman for one of the largest textbook-publishing companies, experienced the frustration of having to establish performance-based goals and personal goals without having any idea what his company's goals were. (In our interview, Jim asked that his last name not be published.) After working for over six days on the company's sales activity report, Jim decided to give up:

> "Here I am, spending all this time working on establishing my own goals when I have no idea whether they are consistent with the direction of the company. I wouldn't be surprised if after I turn this report in they contact me and say that my goal isn't right and that it should be 'such-and-such.' Then I'll say 'Fine!' and I'll get to work. It's just that the whole process is meaningless and extremely time-consuming."

Jim isn't alone in his frustration. In an interview, Janet described one of her first manufacturing jobs at an aerospace company. During the two years she worked in the sheet-metal shop she never knew what she was working on. She knew only that she was to bend a sheet of metal a certain number of degrees before passing it down the line. To this day she still doesn't know what she was helping to build.

Another technique that reduces uncertainty and increases self-satisfaction is to identify the characteristics of the work that indicate its good or bad quality. Employees should get feedback from the work itself; that way they can satisfy certain of their expectations themselves.

Consider the example of a floor sweeper interviewed at a Florida aerospace company. He takes his work seriously. His dedication to doing good work stems from the way his supervisor explained the work to him. The supervisor told him that when he keeps the floor clean he is helping to ensure that the space shuttle and its crew will take off, perform their mission, and return safely. He understands that keeping the floor clean helps minimize the possibility of foreign particles getting into the spaceship and causing a malfunction. He also feels good when he finishes, because the supervisor pointed out certain things to look for that let him judge for himself that he has done a good job.

Management Practice 7: Build the Expectations Package

Definition

An expectations package is a negotiated agreement between a manager and his or her subordinate in which the important expectations of each have been expressed and essentially harmonized. To create conditions for exceptional performance, the manager and an employee need to discuss and define the job in relation to organizational goals. Both need to review what will be done, when it will be done, how it will be done, and with what results.

Two further terms must be defined for use in the discussion of making expectations explicit and congruent. A "need" is some condition, group of resources, or characteristic of the internal environment that is essential to the performance of the work. Needs further refer to essential elements that must be derived from the job to permit enthusiastic involvement and optimal performance.

A "want" is an element in the workplace that is not essential to acceptable performance but enhances an employee's ability to do the job and increases the satisfaction he or she derives from the work. Both the employee and the manager must gain as thorough an understanding as possible of the needs and wants of the other.

Differentiation

The expectations package specifically clears the way for exceptional performance in the high-tech environment. Its chief goal is to allow employees and managers to perform well when passionate personal involvement and a very high level of effort and innovation are needed.

Thus the use of the expectations package is essential in high-tech and somewhat less essential in traditional industries. The major goal of the expectations package is to free employees to make a creative contribution. Negotiated expectations are essential to creative performance. They are not as necessary in traditional work situations where caution and precise adherence to standard procedures are given more importance. But this is not to say that building an expectations package could not enhance performance even in these situations.

Importance

This is the point at which the expectations of employees and managers are actually made explicit and discussed to resolution. Of all the management practices proposed by the high-tech managers, this one is probably most closely related to establishing a climate in which subordinates motivate themselves. If the expectations of employees and managers are not brought into the open, discussed, and negotiated, serious obstacles are likely to remain. Exceptional performance is not apt to be achieved without developing an expectations package related to each project to be undertaken.

How to Do It

After using the exceptional performance management practices for a time, both managers and subordinates will understand that at the beginning of each assignment both parties should be prepared to reveal and discuss their wants and needs related to the project. With continuing work assignments that do not include distinct projects with beginnings and ends, managers and employees should renegotiate their expectation agreements periodically.

The goal of the negotiations is to enable employees to participate in a buy-in of the work assignment. By the term "buy-in" I mean an event in which both parties give something with the expectation of receiving something in return. From the employee's point of view, the buy-in essentially consists of offering exceptional work in return for the important satisfactions that can be derived from the work.

To create this situation, you as manager must manage the buy-in effectively. The buy-in has three components.

1. Discuss the job itself An important part of building an expectations package is arriving at a mutual understanding of the job. Communicate the tasks and goals and verify that your employee understands the assignment. After explaining the work, you must ask the employee to express his or her understanding of the job. When the employee's perception of the work differs from what you meant to communicate, explain further. Keep at it until you both have the same conception of what the assignment entails.

2. Define the worth of the work in relation to the organization's vision Explore with the employee the true importance of the work. The chief goal is to express the expectations of the manager and the overall organization. These expectations relate to the achievement of the company's goals and vision. Discuss this until it is clear the employee understands the contribution he or she will make to the organization's vision.

3. Explore expectations that need to be fulfilled or that may interfere with progress if they are left unmet This is the payoff in creating a climate of motivation. Improperly managed, this is also the point at which exceptional performance begins to be interferred with by the obstacles of unsatisfied expectations.

Expectations of both parties should be explored. The manager may have implicit expectations. He or she will typically expect these from an employee, whether the manager states the expectations or not:

- I want an employee to be completely honest.
- If an employee disagrees with me I want him or her to let me know.
- I want employees not to wait until the last minute to produce work.
- I expect the job to get done even if it means working overtime.
- I want the employee to take initiative and think for himself or herself.
- I will be fair to the employee if the employee is fair to me.
- I expect employees to give me full status reports so I never get last-minute surprises.
- I expect employees to see me first before taking action that might create controversy inside or outside the organization.

- I expect neatness and courtesy from all employees.

Common areas of implicit expectations and desires of employees are these:

- My boss will have my best interests at heart.
- My boss will give me feedback.
- My boss will assign me to projects that are on the leading edge.
- My boss will never criticize me in front of others.
- My boss will help set priorities for my work.
- My boss will know how much work I can handle without feeling too much stress or turning out substandard performance.
- My boss will know that I want to advance in the organization and will help me do this.
- My boss will try to provide the degree of visibility I desire for the goals I achieve.
- My boss will give me work that uses a wide range of my abilities.
- My boss knows that I enjoy travel and can do better work when face to face with other people involved in the work.
- My boss knows that I do best when I am given the autonomy to manage my own work.

Your job, just as the job of your subordinate, is to explore all of these issues and others that arise. Hammer out an agreement about how the two of you will work together. This is one of the most important things you can do to make your operating unit capable of exceptional performance.

During his interview, Larry Biggs of Public Service Electric and Gas (New Jersey) said, "Establishing this understanding can be likened to a union-management negotiation. Both sides know what they want and what they are willing to settle for before the negotiations begin." It's up to management to take the initiative and open up the discussion. As both parties share common interests and experiences, they also work toward modifying expectations until agreements can be reached.

Sometimes it can be a legitimate criticism of management to say that they follow the Golden Rule. "Do unto others as you would have them do unto you" is not a good management rule. You must give others *what they need and want* even if it is completely different from what you would want in the same circumstances.

Getting Started

The most important thing you can do to get started is to make the effort to understand your own implicit expectations about the people who work for you. If you want something from employees, be clear about what it is and tell them! They may negotiate with you and in the exchange get you to accept something else. This modification of expectations is a legitimate part of the process. Under any circumstances, both sides should begin by introspection. If you don't know what you want, you are very unlikely to get it.

8 ACHIEVING EXCEPTIONAL MANAGERIAL PERFORMANCE
Managing Events and Influencing Others

In the mission control room, hardware technicians, software specialists, and mission control managers were raptly attending to their communications terminals. Next to the mission director was seated a man who was serving as neither a technician nor a NASA manager. It was Stephen Schwartz, president of Satellite Business Systems (SBS), IBM's subsidiary that sells long-distance communications services to corporations and government agencies. The space shuttle *Discovery* on this launch was delivering one of SBS's communications satellites to its position in space.

What was Schwartz doing there? Beyond the practical reason of being prepared to make decisions about the satellite if something should go wrong, Schwartz's presence was symbolic of a broad business strategy of

IBM's. His prominent work station in the mission control room told everyone—IBM employees and competitors alike—that IBM was a heavy hitter in the booming communications market.

IBM was at that time sending other high-visibility signals of their serious interest in this market. The company had just spent $300 million for a minority interest in Rolm Corporation, a manufacturer of telecommunications equipment. IBM had greatly increased its capital participation in SBS. The company had further joined forces with British Telecom to market computer network communications products; it had reached agreement with Mitsubishi to develop and market products to Japan's national telephone system, NTT.

All of these actions by the giant computer-maker could not be coincidence. There was a grand scheme behind the moves, a vision, if you will, of deep involvement with telecommunications in the coming years.

Two factors appear to underlie the new strategy:

1. The utility of large mainframe computers—IBM's bread-and-butter product line—is greatly increased by having a convenient, reasonably priced means of communicating with outside users and other computers. By establishing a standard in the communications market, IBM could increase sales of its mainframes and its personal computers, which serve as convenient user terminals.
2. IBM anticipates a major struggle with AT&T in the computer and communications markets. This communications giant has recently won the right to compete

> directly in the computer market. AT&T's intentions could hardly be clearer. AT&T managers have directly stated their plans to be near the top of the computer mainframe business by 1990. IBM is preparing now for the strenuous competition such a large company can create. Despite AT&T's relative lack of marketing experience, the company will be a powerful competitor because of its truly outstanding research and development capability.

IBM clearly has a major strategy for competing in the emerging huge communications market.

What's going on here? How will the long-term implementation of this strategy help IBM? Certainly it has a chance of increasing the size and profitability of IBM's operations. It may directly increase sales of products and services.

According to the observations of the successful high-tech managers I surveyed, there is another benefit to the company that possibly surpasses the obvious financial one: broad strategies, properly communicated to managers and employees, will help create an atmosphere that encourages exceptional performance.

The grand schemes of IBM and many other successful companies give members of the corporation the feeling that they are part of a potent and aggressive organization. Employees feel inspired and motivated by their identification with the power and effectiveness of the organization.

Can You Motivate and Control Employees?

That's an easy question. The answer is "No!" Listen to what interviewee Bill Lamb has to say. Lamb is a project manager for the Power and Heavy Civil Division of Acres American.

> "People aren't managed. They can't be controlled like robots. They are individuals. You don't tell them to 'do this' or formally lay out things to do and approach them by saying 'do what I say' . . . . That's more of an assembly-line approach. Instead, the goal is to direct your people in such a way that they will want to do what you want. After all, the people who work for us enjoy their work. As a manager I have this commonality to work with. The key is to capitalize on it."

Bill Cobb, an organizational development consultant at the corporate offices of Ford Motor Company who was interviewed for this study, elaborates on the same idea:

> "The organization, the project at hand, and the leaders need to create the environment that will enable people to motivate themselves. This is one of the outcomes we hope to achieve with the employee involvement groups we have implemented throughout Ford Motor Company operations. Giving people an opportunity to participate and be involved in the work process, we hope, will help to create the kind of environment where people can get excited and enthusiastic about what they do. Only *they do it to themselves.*"

What is the message of these revealing statements? It is that managers neither manage nor motivate people.

Managing Events

If not manage and motivate people, what does a manager do? A manager manages events or creates situations to *influence* people in the pursuit of desired results. Managing events leads to exceptional performance; properly devised events provide opportunities for people to become personally engaged in accepting the challenges of the enterprise.

An *event* is an existing or created situation in which conditions exist for individual employees to take full responsibility for the creation of their tasks. This ability to participate lets employees and managers exercise their own interests, skills, and motivations. They can arrange the work so it is satisfying, challenging, and interesting.

Consider two approaches to a corporate goal:

- Meijer had a goal of introducing personal computer use by corporate managers at all levels. The company president authorized walk-in computer centers in the company's home office. Computers were installed in each of the centers along with a basic instructional manual and a note: "Anyone who would like to learn, please call ______." Within a year almost every manager wanted the training.
- A hypothetical company has the same goal of introducing personal computer use. The top managers of this company send memos to each department informing them of the new procedure. The memo says that, in order to do their jobs, managers must learn to operate the computers within a year. Formal training for each manager has already been scheduled. After a year, all the equipment has been installed but use is only a small fraction of what was desired.

There is a clear difference between these two methods. In the first example, company management created an event. They developed a situation in which employees could choose on their own to learn about the computers. Employees were free to create the conditions of their own adoption of the computers. They had choices: they could experiment on their own in the computer centers; they could pursue the formal training offered by the company; or, if they chose to, they could ignore the computers altogether.

In the second example, employees had no control over the conditions of adopting the computers. Their only alternative was to accept the equipment and undergo the formal training that had been scheduled for them. Under these circumstances, employees often revert to elaborate subterfuge to give the appearance of complying with directives. Pretending to comply while actually doing the opposite may even become an informal game.

When a manager devises events, he or she must at the same time manage the risks and resources related to the participative actions of subordinates. Consider this situation as an example.

> An electronics technician has been assigned to three project managers to make wire-wrap prototypes for applications testing. All three of the projects are on a tight deadline but at the same time require considerable precision in work results. As a result, the technician is under tremendous pressure. She is lacking the resource of time.

The technician's functional manager has the responsibility to intervene in a situation such as this. He or she must work with the project managers and reach an agreement for the technician's services. The agreement must consciously address the risk associated with delay of the projects versus the stress and difficulty caused for the technician. If the agreement is negotiated properly, the technician will be able to have significant control over her own contribution to the projects.

When devising and executing events, the delegating manager must also manage the relationships of trust and confidence that exist among managers in different specialties and at different levels. In our interview, Joan Lefkowitz, director of human resources at Burger King's corporate headquarters, presented this view.

> "[The management of trust] is the fuel that builds commitment toward getting the job done and reaching the end results. When there is a lack of respect and trust in the workplace, it is likely that you will observe people saying things like 'We spend 75 percent of our time and energy covering up our back ends and 25 percent on the job at hand.' Better pay attention to the dancing, not the music."

How does an event user manage trust and respect? There are really only two steps:

1. Negotiate an agreement with a subordinate, allowing the subordinate to play an important role in deciding the conditions of work. The subordinate should have the opportunity to make major contributions about how the work will be done. Autonomy breeds self-respect and respect for superiors.
2. Remember the agreement exactly as made and provide the conditions and resources needed to fulfill it. Absolute honesty and

skill at working the system to provide resources are the key requirements.

Types of Events

Events are conveniently considered to be of three types: closed, open, and final. In addition, managers must consider subevents to support events.

Closed Events Some goals may be achieved through a single event with rather predictable outcomes. In this case, only a single situation need be established to reach the current goal. For example, a manager at a California aerospace company who wishes to remain anonymous attempted to get a percentage increase in pay for his subordinates at merit-raise time. His immediate boss refused the request. But this manager knew his people well enough to recognize the severe consequences on morale and turnover of this decision. He decided to go over the boss's head and arranged a meeting with top management. After discussing the issue at length, top managers approved the increase he had requested.

This is a single, self-contained event. The goal was achieved with a single action—the meeting with top management. Other events may be needed in the future to deal with the erosion of trust between the manager and his immediate superior. Such further efforts, however, are not directly related to the goal of gaining the raise for the manager's subordinates.

Open Events Open events are members of a planned series of events that a manager creates to achieve a larger goal. After the completion of an open event, alternative situations may be set up to accomplish a further step toward the overall goal. Open events are links in a chain that reaches toward a goal. Closed events (if they succeed) achieve the goal through the staging of the single event. In a sense, the event becomes an end in itself.

A simple example will illustrate the function of open events. An Indiana supplier of auto parts to the Big Three was in the process of purchasing a training program in stress management that I had designed. The decision to use my services was easily reached by the company president and vice president over a casual business lunch. However, company management was faced with the challenge of getting all managers—especially managers who were already suffering from symptoms of stress—to take part in the training. Top management considered but rejected the idea of simply ordering everyone to attend, judging that this approach was inconsistent with the company's loosely structured environment. They decided that attendance must be voluntary.

To generate internal enthusiasm for the program, they arranged to have me interview a select group of managers. Managers to be interviewed were chosen primarily for how much influence they had within the group. One manager, Eric, was chosen for me to interview with the specific belief that if he was sold on the program, he could persuade Tom, a resistant manager, to attend.

To carry out their plan, company management arranged a meeting with Eric. The meeting concerned another issue that needed to be dealt with, but the pressing concern of the top managers was getting Eric to influence Tom to take part in stress training. Casually, at the end of the meeting, they introduced the subject, and Eric left assuring them he would get Tom to attend.

There is clearly a chain of events in this example. The original meeting was the first event. In the meeting it was decided to present the stress-management program and to take certain steps to get managers to attend. This first event pointed the way to other events: the meeting with Eric, the interviews with selected managers, a purchasing contract to the consultant, and finally the beginning of the stress program itself.

The achievement of a goal—in this case the presentation of the stress program—is itself an event. Since it is the last event in a strategic or tactical series, it is called the *final event.*

Event Management in the High-Tech Environment

Event management is not new, although this book is new in its focus on events as a distinct management entity. The emerging importance of event management accompanies the increasing need for exceptional performance in high-tech companies and departments. As traditional firms become more and more imbued with the demands of the high-tech environment, their need for event management will also increase.

In a traditional firm, managers are more likely to use direct controls and authoritarian approaches with clearly defined procedures for getting work done. To a great extent, efficiency can be obtained by strictly following the procedures and standards. It is less essential to get employees and managers to emotionally buy in to the tasks and goals of their department. Exceptional performance and a high level of innovation can even be obstacles to success when production efficiency is the main corporate goal.

However, when

1. your organization's people seek the option of being more autonomous,

2. competition is extreme and growing ever more threatening,
3. product life cycles are shorter,
4. risks are greater,
5. new technology is continually introduced, and
6. many employees must function without the aid of long-tested standard procedures,

it is essential for people to be truly committed to their work and consistently able to apply their greatest capabilities to it.

Management of events can contribute tremendously to this dedication, creativity, and persistence. Event management works because it leaves employees free to apply their particular blend of skills, interests, and personality to the goal at hand.

Characteristics of Events

Event management will work only if the originator of events can endow them with certain key features. The more your events reflect all of the following characteristics, the more likely they are to succeed.

- Events should be observable. Having a meeting is observable; thinking about a problem is not. Painting yellow aisle markers on the shop floor is observable; fantasizing about making a complaint to the safety engineers is not.
- Events should focus on achieving a particular result within a definite period of time. If a meeting is the most effective route to the results you want, determine the time needed and who should attend. Then take action.
- Events must be logical and achievable.

Let's say that you plan a series of events to reduce waste material in the production process. You must make sure that the goals are achievable given the tolerance of equipment and design of the process. No matter how dedicated and creative employees are, they will not be able to meet tolerances that are beyond the capability of production machinery. Such a goal destroys your credibility and can actually produce greater waste.

- Events should have goals that team members consider worthwhile. You are asking your people to make a personal, emotional commitment to organization goals—too much to ask if the goals seem trivial or counterproductive. There are two possible situations. First, the employees may be right; perhaps the goals are trivial or misguided. If so, change the goals. Second, the employees may not recognize the importance of your goals because they are not well enough informed about the organization as a whole. Overcome this by providing continual reorientation. Explain fully and accurately how the goals contribute to overall strategic success.

Conscious planning of events is an important management tool. The idea of events provides a good conceptual framework for handling the active parts of the management job. The following five practices were followed by the successful high-tech managers in my survey. These practices, put into effect on the job, can provide good approaches to getting your people strongly engaged in the process of meeting corporate and departmental goals.

Management Practice 8: Develop and Communicate a Broad Vision

Definition

The vision of a manager answers the question "What is our company trying to do and how shall we go about doing it?" The vision is an all-encompassing view of the future strategy of the company. In developing the vision, top managers answer questions such as these:

1. What markets will we compete in?
2. What will be our characteristic response to intense competition?
3. What product lines will we maintain?
4. What will we do with profits generated?
5. How much will our company grow—through retained earnings, through further capitalization, through merger and diversification?
6. What will be our policies related to employees' rights, to customers, and to suppliers?

These issues and others like them are traditionally an integral part of planning the strategic position of the firm. They are the central content of the top management job. The vision described in this management practice includes all of these factors but has a further particular meaning.

The vision includes a mental and emotional picture of the way individuals and groups will work together to reach goals. The success of the vision depends on managers' ability to communicate the enthusiastic "one for all and all for one" thinking that underlies exceptional performance. The vision thus includes strategic plans and goals but also reflects the managers' perceived need for deep emotional commitment to the activities and aims of the organization.

Differentiation

The survey managers agreed that the vision in a traditional setting will largely emphasize matters such as market share, advertising policy, pricing policy, and administrative routines. Even the most traditional companies usually have goals related to future growth, diversification, greatly improved productivity, elimination of wasted funds and materials, and the achievement of transcendentally efficient production. In traditional firms, however, greatest stress is placed on the month-to-month approach to markets, production output, and employee performance and compensation.

High-tech firms obviously deal with nuts-and-bolts measures of productivity and production control as well. These matters do not receive the greatest emphasis, however. The important strategic factor involves a mental image of groups of employees and managers passionately engaging themselves in the firm to reach challenging goals. It is the image of striving, of exceptional energy bringing exceptional results, of minds crackling to yield creative solutions that most characterizes the vision in a high-tech environment.

Importance

If you as a manager are going to ask and expect people to give passionate dedication to the goals of the organization, you had better know what those goals are and be sure the goals are worthy of engaging people personally. This is an important aim of the strategic planning process: to derive a view of the company in its environment that is worth getting excited about.

Further, your people are far more likely to devote their best abilities to the company if they believe it is likely to be successful. Strategic planning is essential to long-term success. Such planning is therefore a desirable end in itself, even ignoring its motivational effect on employees and managers.

How to Do It

The question of how to carry on strategic planning has been dealt with exhaustively in other works. An important point that deserves further emphasis is the need for a truly broad vision that will inspire employees. My numerous interviews with middle managers, upper level managers, and CEOs yielded a single common response: *develop a broad vision by talking to other people.*

Discuss the company's or department's long-range goals with everyone who conceivably will take part in or be affected by the efforts to reach those goals. This means, first of all, to include your subordinates in strategic planning. You are going to ask these people to dedicate themselves to meeting the goals; they clearly deserve a voice in setting those goals. The consultation with subordinates is an event in the sense described above. By instituting the discussion, you have created a situation under which employees can participate in setting the conditions of their own work.

Talk to peers. Their outside view of your operations can be extremely revealing. Peers also will have control of many of the resources your subordinates will need to get the work done. Involving peers in your operating unit's strategic planning may well gain their support and free up some of the resources you will need.

Talk to as many people as possible in the hierarchy above your position. To work, strategic planning must involve the whole organization. Your superiors will probably have set up some formal planning procedures. Come to understand these. Find ways to get your organizational superiors fully to communicate overall strategies. Only this thorough understanding of the corporate direction will allow you to communicate a broad vision to subordinates.

Getting Started

The rules for getting started in developing and communicating a broad vision are simple. The hardest part is simply doing the leg work that creates strategies that are likely to succeed. Here's a way to do it:

- ○ Review current plans for your company and operating unit. Are they likely to inspire people to produce exceptional performance? Do they make a call to human pride and dedication—or are they merely numbers that employees are meant to meet or exceed? If the strategies fail to meet these criteria, note in writing the areas that need to be strengthened.
- ○ Talk with subordinates, peers, and superiors. Make it clear that you want ideas and information related to a vision that will engage people emotionally.
- ○ Review in your mind whether you truly understand and identify with the major strategies of the company. You will probably find areas you understand less or identify with less. Try to clear these up by talking with others again.
- ○ Simplify and capsulize the vision as much as possible without sacrificing meaning. This summarization should suggest sym-

bols and slogans as discussed in chapter 6.

- Use the slogans and symbols, as well as a more detailed description of corporate and departmental goals, to *communicate the vision.*

MANAGEMENT PRACTICE 9: FOCUS SHARPLY ON WHAT IS IMPORTANT

Definition

This concept is two-pronged. Focusing on what is important has two aspects that are, for practical purposes, inseparable: What will we do? How will we do it?

Strategic planning should already have substantially answered the first question. Your company will attempt to meet strategic goals. The key element now is to focus on those goals unwaveringly while work is being carried out. Focusing prevents the common failing of becoming so tied up in the details of events and processes that the overall vision is lost. It is partly the communication of the vision that makes exceptional performance possible; it is a grave error to lose sight of this during operations.

The answer to the second question depends partly on simple operations planning. You and your people probably have the major skills for carrying out this planning.

There is, however, a second aspect of "How will we do it?" that is often given less prominent attention than it deserves. The greater question is "Who are the people who have to move to get this plan accomplished?" Who has the power to authorize it? Who has the resources our people will need? Which of our own people are the most capable of spearheading the work and performing excellently in its completion? Focusing on what's important answers all of these questions.

Differentiation

Management must attempt to focus sharply on the company elements that produce the greatest reward because only certain high-priority items truly deserve a manager's attention. Traditional managers typically follow the 80/20 dictum, which holds that 80 percent of the importance of a manager's work resides in only 20 percent of the tasks he or she is responsible for. If that key 20 percent of the job can be carried out well, 80 percent of the job content will be well managed.

Many high-tech managers interviewed stressed another way of focusing on what's important. The successful high-tech managers measure the importance of tasks and decisions according to how much leverage each will provide. Remember that the truly central goal is to influence subordinates to achieve exceptional performance. Focusing on what's important then subtly shifts toward creating the events that will lead others to outstanding dedication and effort. From this point of view, it is more important to create events that allow subordinates to work with zeal than it is to analyze exhaustively a production variance report. Under any circumstances, someone should closely review the report. Under the approach advocated by the high-tech managers, a subordinate should do it.

In high-tech enterprises, there is a special temptation to lose sight of what's most important. Much of the technical innovation and process management of high-tech companies and departments is quite absorbing. It is easy to become mired in the technical aspects of a problem and lose sight of the really important issues of reaching goals. Thus, it is especially important to make a conscious effort to stay properly focused.

Importance

A person in the position labeled "Manager" is truly a manager only if he or she both governs the direction of work and accomplishes the work through the efforts of other people. Clearly, focusing on what's important is central to the management job. It fulfills two purposes:

- △ It keeps the efforts of the operating unit aimed toward the most important strategic goals of the department and company.
- △ It greatly enhances the likelihood that employees and managers will be able to apply to assigned work their own motivation and excitement to achieve.

How to Do It

Developing the focus on important matters requires a clarifying review of strategic directions along with communication, communication, and more communication. The difficulty is that you must yourself maintain the proper focus and at the same time see that your subordinates and key superiors are properly focused. Trying to do this will combine activities that reinforce the broad vision with some fairly detailed efforts to bring out the critical elements of strategic plans.

1. Identify the key people and gain their support for accomplishing the most important goals The key people are those who have the power and control the resources you need. Create events that will influence or give your people support. First, identify the personal style of these key people. For convenience, we may divide these styles into three categories: action, conceptual, and uncommitted.

- Action people are risk-takers and become impatient when discussions grow too far-ranging and speculative. Action people want decisions and an action plan even when many unknowns remain.
- Conceptual people tend to be analytical. They want to understand the important forces in a situation even if gaining this understanding delays action.
- Liquid people respond primarily to the most immediate situation. Their excitement about a project often disappears as soon as the proposer leaves the room. They may appear enthusiastic about a course of action, but they often delay decisions to avoid committing themselves. Thus they avoid making waves.

Suit the events you plan to these styles. Action people want highly summarized information along with concrete, numerical projections of the effects of a decision. Conceptual people want detailed information about the many inside and outside conditions that affect a decision. Uncommitted people want the information couched in a manner that will protect their security. They want an easy choice set up so they don't need to rock the boat. Having other influential managers check off your proposal before presenting it to an uncommitted person is usually helpful.

2. Suit events to the personal goals of subordinates and peers Contrary to belief, many people in your organization don't wish to be stars. Many prefer a low-profile working position without the threats or the rewards of a high-visibility, "fishbowl" position. Manage events so your subordinates can plan their work conditions to protect their personal preferences.

3. Use symbols Symbols have already been discussed in chapter 6. In that context, the important role of symbols was seen to be in creating and reinforcing pride and commitment. Symbols are also used to elicit enthusiastic response to specific, important tasks and goals.

Remember that a symbol is an image or slogan that represents something else through association, resemblance, or convention. Symbols can directly influence large groups of people; properly used, symbols produce active results in daily operations. When carefully established, they can help create a widespread sense of value in an organization. A part of the desired value system, of course, is devoting exceptional performance toward meeting organization goals.

Symbols that are directly related to building values in the organization might be called *image symbols*. Image symbols should not be confused with power symbols. *Power symbols* are representations of status and authority within the organization. Common examples are the size of a person's office, its number of windows, the quality and size of office furniture, office location, and whether or not the person has a personal secretary.

Image symbols are representations expressly and strategically created to inspire people to work exceptionally well to achieve a desired end result. Such symbols relate to the organization and its work rather than to the status position of organization members.

Consider some examples of image symbols in society at large and in business organizations:

- Lyndon B. Johnson's "War on Poverty" slogan summarized myriad social programs. The symbol of waging war helped unify many diverse groups behind the programs.
- Ronald Reagan used a similar slogan—"A

New Beginning in 1984"—to capitalize on the common feeling that the quality of American life and America's international influence were waning. The slogan again provided a rallying point for many diverse groups.

- Churchill represented the indomitable British spirit by his often-used "V for victory" gesture. This gesture spoke to both the British people and the warring Axis powers.
- The letters IBM stand for a constellation of products, employee characteristics, and customer relations. Anyone hearing "IBM" immediately thinks of computers; technicians and salespeople in three-piece suits; expensive but excellent service; secretive strategic planning; and all the other characteristics that through the years have come to be associated with "Big Blue." IBM at one time widely disseminated the motto, "Think." This is also an image symbol. It signaled to both employees and customers that IBM was the sort of steady, stable company that thought before it acted and analyzed all factors before committing.
- GM, Coca-Cola, DuPont, and scores of other companies have symbols that are widely recognized to stand for the characteristics of the company.
- Avis built an extremely successful marketing and internal development program around the idea that they "try harder" because of being number two in the rent-a-car business.

For symbols to work in influencing people, they must have certain characteristics:

1. Symbols must be simple and easily recognized. Consider the great range of ideas represented by IBM's single word, "Think."
2. Symbols must be coupled with an entire program related to the goal. If a company adopts the quality-control slogan, "Zero Defects," the slogan will be ineffective unless the capability for zero defects is built into the system. A constellation of manufacturing procedures, exhaustive quality testing, employee rewards, and standard-setting would be needed to make the slogan work.
3. Symbols must be realistic. If a company adopts the slogan, "We're the best," employees are not going to be influenced if the company really isn't best at all. The slogan would succeed only if tied in with a major improvement campaign that presented being the best as a concrete, achievable goal.

Getting Started

The steps for getting started are clear:

- Use your strategic plans to decide what is really most important. Pick a reasonable number of points. Subordinates' morale is damaged if you say that everything is equally important and has first priority.
- Devise events related to the accomplishment of these important points. If quality must be improved, work out a series of open events leading to quality-related final events.
- Identify the important people who have

the authority or influence to affect the achievement of your major goals. Plan events to gain their support. Suit the nature of the events to the personal style of these important people.

- ○ Identify the key subordinates who will contribute most to the achievement of the important goals. Plan events that will let the subordinates approach the work in a way that suits their personal goals in the organization.
- ○ Plan to use symbols to support the program and to inspire subordinates to produce exceptional performance. Subordinates' participation in designing their own work, coupled with powerful image symbols, will work wonders.

Management Practice 10: Act Decisively to Empower Others

Definition

Empowering others is the act of authorizing a task to be completed by someone other than the empowering manager. Two conditions must be met:

1. The person empowered, not the empowering manager, will receive the glory of success.
2. The person empowered will have personal control of the resources needed or, alternatively, will be authorized to take action to gain the resources himself.

The empowering manager is, in effect, sharing the commitment with others and giving them credit for achieving the goal. By empowering others a manager is practicing *indirect management.* He or she sets up a series of conditions under which individual employees in a team are responsible for the creation of their own work. Genuinely empowering others aids the achievement of exceptional performance. When others are allowed to set the quantity and quality of their own work, they typically set goals that are higher than expected.

When a manager empowers others, he or she influences rather than directs them. Once appropriate situations are established, subordinates will be led by the features of the situation to provide their own energy, creativity, and skills. Individuals become engaged in the project, feeling a sense of meaning in and ownership of the work and goals.

Empowering is the next step beyond Management Practice 2: Practice letting go. Letting go involves helping your people learn and gain self-confidence; it is an essential preliminary to empowering. Once they have learned the needed skills and gained a high level of self-confidence, they are ready to be empowered.

Two types of empowering can be distinguished. In *directive empowering*, the manager initiates the activity. He or she approaches others to get them involved in a project and create enthusiasm for signing up. The manager initiates situations in which others can find ways to satisfy themselves by working toward a goal set by others.

In *receptive empowering*, the goals and activities originate with subordinates or peers. In creating their own jobs, subordinates propose projects and activities in which they would like to engage.

Differentiation

Empowering others is a key response to the difficulties of the high-tech environment. There are several reasons for this.

- Strong competition demands tough management responses. Managers can almost always be tougher when their own interests are involved than when they are merely representing the interests of another person or organization.
- High-tech enterprise demands very high levels of expertise in a variety of subject areas. A manager's subordinates will often have greater expertise in some areas than the manager does. It is only through empowerment that this expertise can be fully tapped.
- Typically, high-tech people are better educated and inherently more self-motivated and autonomous than traditional employees. Many, left to their own devices, will identify themselves as readily with their profession as with the company they work for. Empowerment breeds company loyalty and at the same time satisfies employees' needs for autonomy.

Full empowerment may be less effective in traditional companies for which maximum efficiency is the overriding concern. To achieve peak efficiency where relatively little innovation is demanded, it is often more important to follow carefully analyzed procedures than to have maximum commitment and creative involvement with tasks.

Importance

Empowerment is important because it adds a major feature to simple delegation. It automatically allows subordinates to become personally involved with the higher-level goals of the company.

Empowerment differs from delegation, which gives fairly specific task assignments. Empowerment creates a climate of passionate engagement. Consider the example of a football team. If the coach calls the plays, he is delegating. If the coach creates the climate in which the quarterback has the savvy and commitment to call as well as execute plays, the coach has empowered the quarterback.

This step beyond simple delegation is the key. When you rely absolutely on the technical and creative skills of your subordinates, you want them to have the ability and confidence to do the job.

How to Do It

Consider these guidelines first:

- ☐ You must be self-confident enough to be tolerant of mistakes. If you are terribly upset when your people fail, it is unlikely that you will ever feel comfortable with true empowerment. When others are allowed to create their own work situations, mistakes are bound to happen. Further, consider whether you are self-confident enough to give all the credit to whom it belongs—the person(s) empowered. The boss of a data-processing manager for ABC in New York had this confidence. The manager told me, "My boss never comes around to check up on us and see how we're doing. The only time we see him prior to the completion of the project is if priorities have changed or an emergency situation has arisen."
- ☐ The necessary elements and agreements for empowerment have to be in place. Determining whether they are is called "parameter identification." Make sure you have (a) a clear understanding of the need for a project and of the desired end result;

(b) a set of criteria for judging success; (c) agreement on what constitutes the final output; and (d) a willingness to accept the risks involved.

- You must be willing to let go. This of course requires you to trust the people you are empowering. Check this trust by looking for these characteristics in the subordinates you empower: (a) willingness and ability to do what they say they will do; (b) keeping you informed voluntarily; (c) making themselves available to you; (d) proper attitude, admitting errors and recognizing and accepting responsibility; (e) possession of good self-evaluative skills; (f) ability to appropriately use money, time, space, personnel, and other resources; and (g) potential or demonstrated expertise in the area under consideration. Letting go of control is difficult. Jim Gorman, training director for the S.E.C., said while being interviewed, "It's human nature to want to control. However, it's also human nature to want to be free; balancing the two is the challenge."
- Use progressive event-building to bring your subordinates to truly exceptional performance. Decide what needs to be done before a subordinate can be turned loose on a project. Establish progressive events that will provide information and background; resources; learning; experience; assistance; a model to follow; and proper self-confidence, pride, and commitment. Carry out the events and judge the extent to which the subordinate is emotionally

buying in to the project. More events may be called for if the level of commitment seems insufficient.

Finally, stay out of the way. Only if people do not consistently operate within the established guidelines and parameters of their agreement should you step in. You also need to intervene if the end result needs to be changed. Otherwise, let the empowered subordinate run the show and get the credit.

Getting Started

First, judge the readiness of your subordinates to be empowered, using the criteria given above. If adjustments need to be made, and they probably will, consider preliminary events to build skill and self-confidence.

Begin empowering with a task or project that is important but not absolutely critical. Analyze the check-offs and resources that will be needed. Set up a first event, possibly a meeting at which you review the general characteristics of the project goals. Quite possibly your subordinate(s) will not buy in at this time. But having thought over the situation for a few days, they will likely approach you with a possible way of meeting the goals. These proposals will usually reflect their personal goals, so they should be accepted if possible.

Establish further events to supply resources and more detailed information about the project. Invite communication but do not demand it. Then—stay out of the way. Let the subordinate carry out the agreement you have reached.

If the first occasion of empowerment is successful, you will probably feel confident enough to begin adopting empowerment more and more as the standard way of working in your operating unit. When this time comes, you and your subordinates will be well on the way to exceptional performance.

MANAGEMENT PRACTICE 11: COMMUNICATE A SENSE OF COMPETITIVE URGENCY

Definition

The feeling of competitive urgency is an internal climate that inspires employees to work to make products, services, and marketing efforts first and best. This sense depends on the perceptions that competition is fierce, that speed counts, and that only the best is good enough. When the sense of competitive urgency exists, the manager has influenced others to see the strategic importance of their tasks; communicating this importance creates the desire to take action and do one's best.

Differentiation

Because of the forces in the high-tech environment—rapid change, intense competition, and the need to be first—the manager who can communicate the immediacy of competitive threat has an edge in product development and marketing. Some companies with whom I have consulted—Burger King, IBM, AT&T, Merrill Lynch, and Executone Business Systems, for example—have created this kind of excitement. The atmosphere brings the best and most from their people.

The difference from the run-of-the-mill office environment is strikingly observable. The pace is faster; energy levels are high. People even walk and talk faster, and their faces are more animated and expressive. Few people waste productive time by gathering in groups to discuss sports or other nonproductive topics. Employees feel there is no question whether they would work unpaid overtime to complete a project. They feel, "The job has to be done; that's what's most important."

Compared with these employees, people in the ordinary office routine appear to be sleep-walking. The pace is deliberate; focus is lost. There always seems to be plenty of time to get the work done even if it isn't finished on time or up to top quality standards.

The rate of innovation demanded by high-tech enterprise naturally can thrive best with the first group of characteristics. Any company will probably be damaged by an organization-wide culture illustrated by the second set. Traditional firms with rigid routines and practices, however, are far better able to function under the low-energy approach often seen.

Importance

The fact is that in the high-tech world, competition is intense. Consider Storage Technology Corporation, formed in 1969 to compete directly with IBM in the computer-tape and disk-drive market. Sales grew to over $1 billion with a net of over $80 million by 1980. But in 1984, faced with a one-quarter operating loss of $60 million, the company filed for protection under federal bankruptcy laws.

The reasons for the fall were very clear. IBM is a tremendously tough competitor. A series of technical misjudgments by employees and managers hurt. Company founder Jesse I. Aweida was pushing for an excessive growth rate.

Competition is real; it is competition that threatens companies' very existence. An internal staff that doesn't catch fire with the realization that they'd better be first and best doesn't have much of a chance in the high-tech markets.

How to Do It

Managers create the sense of competitive urgency by serving as models, using work deadlines, influencing people to challenge each other, reporting on the competition's activity, and providing energy-rejuvenating events.

1. *Serve as a model* Use personal characteristics, and broadcast enthusiasm and urgency. According to my interviewee Dr. Logan Cockrum, administrator for executive development office of the Comptroller of the Currency, "Individuals who make the best models have a natural sparkle, a kind of charisma that captures people's attention; at the same time it creates in [others] a kind of excitement."

To properly model competitive urgency, you must be well informed and you must maintain a high visibility to your people. Keep your nose to the grindstone. Use every opportunity to convey that the work subordinates are doing is important to the company's competitive position.

2. *Use deadline pressure judiciously* Many people do their best work under pressure. You can use reasonable deadlines to create a sense of pressure if you can realistically tie the deadlines to a competitive need. It is important to explain the reasons behind the deadline; otherwise, hard-to-achieve deadlines will have a negative impact on performance.

3. *Induce people to challenge themselves* Dayton R. Males, Jr., vice president at Booz-Allen and Hamilton, who has been a manager over twelve years, told me during an interview that he uses the chance of promotion as a vehicle to inspire performance. He periodically invites his subordinates to breakfast or lunch. During the conversation, he explains to them the promotional

process. "At that time," he says, "I will interject the idea that those who produce more move ahead. Since I have had informal breakfast or lunch meetings to talk about this idea of personal advancement, I have noticed that people as a whole seem to operate on a higher plane."

4. *Report on what the competition is doing* This seems obvious, but it is often overlooked. It is a mistake to assume that subordinates are as well informed about the competition as you are. Let your people know in detail what challenges other producers are issuing. Never underestimate the possible efficacy of competitive ploys. Your subordinates may be more canny than you think; they will benefit more from an accurate assessment than from over- or underestimating competitive strength.

5. *Use events to renew competitive energy* Use event-management practices to create situations that either directly or indirectly contribute to the feeling of urgency. Properly exploited, these events can at the same time help employees and managers renew their supplies of competitive energy. The events should be brief, with some definite payoff coming almost immediately.

Informal or formal meetings where managers report their successes are a good idea. Stress successes; underplay failure in these contacts. One manager has regular informal breakfast meetings to help disseminate information about a project's progress. There is no agenda, but he serves as a catalyst by leading people to discuss the successes they have been achieving.

Competitive urgency can also be communicated at family picnics, informal gatherings after work hours to shoot the breeze, company cookouts, sports tournaments, golfing meetings, and similar events.

Getting Started

First, make sure you are as aware as you should be of what the competition is doing. Find out from your superiors. Stay on top of your professional reading. Be involved with your trade or professional association.

Second, analyze the competitive importance of the projects and tasks being carried out in your operating unit. Communicate this importance to your staff, formally or informally.

Third, plan the events that you will use to maintain internal competition. Committed people will enjoy their own successes but will strive for even more exceptional performance when internal competition exists.

MANAGEMENT PRACTICE 12: CREATE ACTIVITY AND MOMENTUM

Definition

The essence of this practice is to "make it happen." The exceptional manager depends on his or her subordinates to take initiative and to get the work going through the mill. However, times will always come when stalemates are reached, when the proper approvals cannot be immediately obtained, or when overly close analysis and speculation stop the gears from turning.

That is when the exceptional manager must step in to create purposeful activity and momentum. At this point, the manager must ignore details and push events into motion, concentrating on what's most important in the situation.

Differentiation

In a traditional setting, emphasis will often be put on avoiding wrong decisions rather than on getting the quick action needed in a very innovative environment. When the main goal of operations is greatest efficiency, this approach is reasonable. It is bad practice to rush into programs and enterprises that can threaten the integrity of the company.

The situation is entirely different in the high-tech environment. Innovation is the lifeblood of these companies. Avoiding risks is more dangerous than accepting them. A conservative company that analyzes every issue exhaustively will typically be left in the dust by more aggressive, freer competitors.

Importance

Having the ability to make it happen is important for any manager. As was just pointed out, this is especially true in a high-tech environment where people, ideas, and products need to move quickly to remain cost- and product-competitive. Procrastinating in a fast-paced, competitive environment is not conducive to success. By the time the slow mover is ready to take action it may be too late; another fast-responding company may already have grabbed most of the market.

How to Do It

Creating effective action depends on a good understanding of the internal and external environment. There are several specific factors to deal with.

1. Suit your action to the culture What sort of innovative risks are the top managers of your company most comfortable with? If risk is very threatening—and it is in many companies—try to find a sure-fire supplementary form of the action you wish to take. Press ahead aggressively with that less-threatening event, meanwhile strengthening your arguments for the acceptability of the whole program or campaign.

2. *Make sure that what you make happen is something your superiors want to happen* If projects are being delayed because of a slow approval process, make sure that nothing other than administrative logjams is creating the holdup. Sometimes delays in action actually reflect unstated opposition. Tactful questioning will often reveal this situation.

3. *Always reveal the goal of your actions* One district manager for AT&T Communications has the reputation of being the spark plug in the company. His superiors tolerate the many different approaches he takes to make things happen, whether getting the company in parades, commissioning art work, or sponsoring community guest nights. His superiors are accepting because they know he has the company's goals, objectives, and vision clearly in mind.

4. *Overcome "people problems" with sensitivity* Others in the organization stand to be affected by the actions you take. Plan on this in advance and include overcoming such obstacles as a part of your tactics for creating action. The trick is to find what others' interests are. Ask them and observe their actions. Take their interests seriously and, when necessary, develop compromises when conflicts with your own interests arise.

5. *Accept an **appropriate** level of risk* In my experience in consulting with companies on handling stress that results from accepting risk, I have observed

several approaches that people use in deciding whether to accept a given risk. The first way is to consider the worst case. If the worst thing that can happen as a result of a risk is acceptable, take the risk. Second, establish situations in which there is no way to back down from the consequences of risk. A Chinese war proverb states that if an army is being attacked frontally and on both flanks, the best tactic is to close off the fourth side: people will fight with furious zeal when they know there is no escape. Finally, use an introspective approach. Some people fantasize that they are near the end of their lives looking back over what they have accomplished. Then they ask the question, "Would I be proud of taking this risk?" If so, they go ahead with the risky proposal.

The goal is to take actions for which associated risks are calculatedly in your favor. Whenever you take action—and particularly when you may be stepping outside established procedures—you incur the risk of failure. By examining the odds and by using intuitive skills, you can reduce your risks to a reasonable level. In the final analysis, a manager will take the risk and make something happen when he decides that it's easier to say he's sorry than it is to wait around for a long series of check-offs and approvals.

Getting Started

This is the point at which to review what you have done so far. Look again at what's most important. Decide on the truly critical events that should have moved matters along. Check that the people you identified as most needed for success were correctly chosen. Find who is actually causing the stalemate or lack of action. Assess the likely risks and the worst-case outcome of your personally taking action. If the risks are acceptable, start gearing up.

9 ACHIEVING EXCEPTIONAL MANAGERIAL PERFORMANCE

Facilitating Events by Managing the Process

Consider a simple event that takes place many thousands of times during a typical American workday: a manager faces a problem and wishes to have a meeting with other people involved to share information and get others' ideas about a solution. There are two ways to go about setting up the meeting. The difference between the two approaches illustrates the major subject matter of this chapter.

- Manager A calls each of the people he would like to meet with and asks them when they would be free. After calling key managers several times, he finds a time and date acceptable to most of the managers.
- Manager B sets the meeting time at 11:30 a.m. Thursday. She then calls each of

> the people she would like to attend. After explaining the importance of the problem and the impact it will have on the people involved, she finds that all of the managers except one are able to change their schedules to attend the meeting.

Both of these are legitimate approaches to setting up a meeting. Method 1 usually requires more effort, but it could be expected eventually to succeed in establishing the meeting time.

The two approaches illustrate the difference between event management and process management. In event management, you create the event and then allow and encourage others to buy into it. This is the method used in the second example above. People who are strongly involved in the content of the meeting must themselves make the effort to attend.

In process management, the emphasis is different. Interaction with others in the organization, rather than the conscious establishment of the event, is the major focus. This is illustrated by the first example. Process management consists of using the communications channels, procedures, and policies of the organization in the ongoing operations of the company. In process management, the manager may not be consciously aware of what the events are.

In the first approach described above, it would be typical for Manager A to start with an idea even less definite than setting up a meeting. He might simply know that he had a problem. He would then call others who were involved with the problem and its solution and discuss the situation. Only after several such contacts might the realization emerge that a meeting is needed. This slow emergence of specific events is typical of process management.

Exceptional performers use both the management of events and the management of the process. Under the most desirable circumstances, the two efforts combine to engage others in their work and to create a greater commitment to corporate goals.

Consider a further example. A New England company has engineered an ultrasound device for cleaning small items in the home. The company has never produced a consumer product before; they have concentrated on the industrial market. People on their sales staff in the past have dealt only with industrial buyers who purchase the company's output as components in larger systems. The marketing vice president wants the sales staff to make strenuous efforts to place the product widely with wholesalers and distributors, and directly to a number of the larger chain stores.

Influencing the sales staff to make this outstanding effort demands a long series of interrelated events. Let's look at only one or two.

1. The marketing manager, long before the sales meeting, sets up a demonstration of the device in the company's offices. The display includes an array of household objects covered with grime. Salespeople and others who pass are invited to try out the ultrasound cleaner on the spot. In this instance the decision to place the demonstration booth in the offices constitutes an event. The actual tasks of having the display designed, making signs for it, and placing it in operation are part of the management of the process.
2. The marketing manager also lets everyone on the sales staff take a machine home to try in a real setting. The event is the distribution of the machines to the staff. Pro-

> cess management is needed here, too. It is essential that the sales reps intimately understand the operation of the equipment. This need is met by well-designed instructional and descriptive materials made available to sales reps and other company employees upon request. At this point, there is no formal pushing of the machine on the sales reps. Their own curiosity and drive to succeed lead them to seek the information they need. Further, the marketing manager expects that reps will get a lot of positive feedback on the machine from neighbors who see it in action. This positive response is meant to produce more enthusiasm and commitment in the reps. Arranging the distribution and preparing the instructional and descriptive materials are part of the marketing manager's job of process management.

In this example, events are concrete observable occurrences directly involving the target employees, the sales staff. The process management tasks lie in the background but facilitate the events. Tasks such as providing a physical design for the demonstration booth; arranging to have objects that need cleaning in sufficient supply for the trial runs; and writing, designing, and printing the instructional materials are part of the process. These tasks are invisible to the target group, yet they make the visible events possible. The process consists of all those activities that underlie specifically planned events.

Processes are a means to an end. Processes are the activities that make events possible. It sometimes happens that the process activities themselves are accepted as ends. This never works optimally.

Say, for example, that a company installs participative management programs, such as quality circles or employee involvement groups. A danger is that major emphasis will be put on the program itself and not on the organizational changes that the program is meant to produce. Managers may concentrate almost exclusively on all the process activities needed to bring about the programs. Then the employees meet, express their ideas, listen to others' ideas—and that's the end of it. Their ideas are never reflected in the operation of the company. The final event of producing change and genuine, rewarding involvement is never addressed.

Managing the process obviously is nothing new. It is a keystone of the most traditional administrative approach to management. However, the way processing is used by the exceptional high-tech managers is different. They use it specifically to achieve the events that lead to goals. A more traditional management emphasizes processing as if it constituted the bulk of the management job.

Yet even the exceptional performer recognizes the need for good control of the process. The specific work tasks that employees and managers perform will not necessarily be well performed unless conscious control and support are provided. The exceptional managers influence subordinates to accept responsibility and to bring strong commitment for getting the job done in a way that meets the most important goals. Even when subordinates accept this role, they need your continuing day-to-day support and aid.

Management Practice 13: Sense the Situation

Definition

Situation sensing consists of applying trained, experienced intuition to discover how work is progressing to its end goal. Exceptional performers sense situations to keep abreast of how effectively operations are moving toward implementing the most important points of strategy. Situation sensing is the lively, unstructured, creative element of management. In its minimal form, it is management by walking around.

Using situation sensing requires that a manager be able to assimilate large amounts of seemingly unrelated information from the workplace and, through internal processes, arrive at an overall insight into the level of success of current operations.

The phenomenon of situation sensing can be likened to the closing scenes of the movie *Star Wars*, where Luke Skywalker is about to attack the Death Star. Besides wearing the Jedi helmet on the shield of which he does incredible projections and magnifications of data to help him conquer the enemy, Skywalker is aware that "the Force" is with him. As he is about to approach the enemy, the Force speaks to him. Skywalker removes the shield from the helmet. Instead of relying on logic and information, he lets the Force guide him. The Force manifests itself as heightened intuition, which Skywalker has learned to use—in preference to logic—to give him the strength and insight to overcome the enemy.

Differentiation

In the traditional firm, a highly developed control system is meant to give managers early warning of instances where goals and target amounts are not being met. The production variance report, the budget variance report, and the sales variance report, combined with periodic profit-and-loss statements, do indeed provide information useful for management control. These reports are part of process management in even the most innovative and successful high-tech companies and departments.

The truly exceptional manager, however, uses personal skills to go beyond the numbers of the formal control system. It is quite possible, for example, to find no significant variances in the standard reports, yet to sense that serious problems are brewing with morale, individual involvement, or the accomplishment of non-numerical goals. It is in this area of operations *beyond* quantitative targets that exceptional management produces exceptional results.

Further, situation sensing can be distinguished from the problem-solving process so often emphasized in traditionally managed firms and departments. In high-tech environments, situation sensing is a process of gaining knowledge of *all* current conditions. The information is then digested to reveal obstacles and opportunities. Emphasis is on answering questions—technical as well as managerial. The simple cause-and-effect model frequently offered as the basis of a problem-solving process is inadequate. The variety of forces operating in a truly innovative atmosphere is too great to analyze rationally.

Importance

Situation sensing is like intuitive value added to the control system. In an environment of continual change, lack of long-tested procedures, and doubt about the completeness and accuracy of data, something beyond financial and production control is needed. The need is made even greater by the very results that produce exceptional management.

When people are in control of their own work

conditions, they cannot be completely subjected to a standardized, company-wide reporting and control system. The very message of exceptional performance guarantees this. You are telling your managers, "Do it your way, but do it right and on time." It isn't possible then to require them to follow procedures that perfectly match your quantitative control system. Intuitive situation sensing is the only answer.

How to Do It

Situation sensing follows the model of problem solving, even though it transcends the rational approaches typically associated with the problem-solving process. There are two imperatives of sensing the situation that roughly correspond to the levels of a rational approach to problems:

- Sense a problem situation or opportunity before it happens or in the early stages.
- Sense the features of a situation that will let you solve the problem or exploit the opportunity.

Both of these are needed for exceptional management of the process. Although situation sensing is a holistic skill, it is possible to analyze components of the process. The following four components of situation sensing work almost automatically for those who have developed this skill. The four elements function simultaneously.

1. Have a broad knowledge base The essential feature of the knowledge base is thoroughly knowing the business. The knowledge must include skill and confidence in technical material and an awareness of the characteristics, interests, attitudes, and experience of the people involved in a situation. The National High-Tech Management Survey showed clearly that the majority of exceptional performers considered "people skills" more important to outstanding management than are technical skills. But both are essential. Managers need to include in their knowledge base information and intuitions about the people they are working with.

More and more, successful managers must have outstanding technical skills in the area they supervise. It is not true in high-tech enterprise that "a manager is a manager." A manager good at running data-processing operations is not necessarily capable of managing the manufacture of chips or the servicing of automated production equipment. High-tech operating units present many problems that only technical ability can solve.

According to the survey, technical knowledge is not an end in itself for a manager of a technical area. Rather, the knowledge is chiefly used to interact with others who have great technical expertise. The manager needs to understand a technical problem sufficiently to know whom to call on, to describe the problem and goals, and to understand and implement proposed solutions. Being able to do these things effectively requires a more than superficial skill. Managers need a good general grasp of the technical processes involved in their business.

During my interview with Rick Custer, Manager of Advanced Programs for the Rocketyne Division of Rockwell International in Cleveland, Ohio, he summarized well the questions that a technical manager needs to be able to answer:

- What is being developed in his or her particular field?
- Where does the manager's technical specialty fit into the broad picture of the company and the industry?
- What trends are developing in the technol-

ogy itself, in marketing and distribution ploys, and in the general business environment?

- What are the current terminology and expression styles in the industry?

The sources of this technical knowledge base are the traditional ones: trade and technical journals, professional and trade association meetings and publications, workshops, seminars, refresher courses at universities, in-house training, and informal discussions with others in the field.

Knowledge of the people in the system should grow steadily through day-to-day contacts. Other knowledge should also be kept ready to hand in a personnel file. A manager needs to know people's skill levels, abilities, experience, training—both original and recent—attitudes, aspirations, career plans, and general personal styles. Managers should know the people who work for them as well as they know their fairly close personal friends.

Further, a manager should maintain an accurate mental picture of where each employee stands at the moment. This picture should include progress on current work assignments, social contacts of each employee on the job, general morale, recent successes and failures, and other factors directly related to the work situation.

2. Keep in mind a model of the world A manager's model of the world is a mental concept of what the work system contains and the interrelations within the system that make it function. This model is not a graphic image such as an organization chart. Albert Low, in his book *Zen and Creative Management*, says, "A manager doesn't have an image or picture in his mind when he works. Instead, he has an idea. It is an idea that guides him to his destination or his product."

Defining how the system works can be a stimulating mental exercise. Ask questions: Who works best with whom? To whom do people go when they have a technical problem? A human or managerial problem? Who is improving his or her output? Whose output is declining? Are there cyclical changes in the effectiveness of work in the system? Who is hoarding resources? Who doesn't *have* any resources? What might employees be willing to talk about among themselves that they wouldn't reveal to me?

3. Learn to scan Despite some serious academic study, no one knows for sure what intuition consists of. Superior intuition seems to stem from an outstanding ability to maintain attention and alertness. A sensitivity to the feelings of others, as expressed in their words and gestures, appears to help. A personal slant that accepts ambiguity and risk is a further aid.

Most current thought on intuition agrees that scanning ability helps. Scanning is a cognitive process in which a manager locks on to what is going on around him or her with unusual attentiveness. The process is analogous to scanning the skies with a radar unit or a radio telescope.

The survey managers were definite about the value of scanning as a precursor to intuitive situation sensing. Scanning is especially needed in the high-tech environment, because even relatively minute changes in internal conditions can make a big change in the level of success achieved.

In scanning, the mind is engaged in lateral and associative thinking. A scanning manager is using an innate ability of the mind to absorb raw information from diverse sources in a noncritical and nonjudgmental way.

Managers use different methods to enhance their own scanning abilities. Some suggest focusing on change: look for attitudes and behavior different from what they were yesterday or last week. Brian Burns, president of Employee Benefit Concepts, and an interviewee in the study, listens especially for expressions of frustration in his people. Frustration is a reflection that an obstacle exists. "Of course, this approach," Burns said, "has its premise in the belief that out of adversity there is opportunity. . . . You just have to be patient and work hard enough to find it."

Another approach mentioned to me by Mel Davis, human resources director for the automotive division of Rockwell International, is staying in touch by listening and observing. When walking around the office, the scanning manager notices people talking in small groups. Are they talking about the task at hand or is there much non-work-related discussion? In the flow of work projects, are the people really spending most of their time on productive work or are they bogged down in reporting procedures and other paperwork? In the manufacturing plant, what is the work pace? Who is waiting for materials or equipment set ups? What is the observable energy level? Do people feel an urgency to get back to work after the lunch break?

Scanning should not be limited to observations of the workplace. The exceptional managers expressed a great thirst for knowledge of the world around them. Heightened awareness of forces in the economy, in society, and in the manager's particular subculture of the industry gives him or her more reliable gut reactions. It is practically impossible to know too much about the environment that surrounds a company.

4. Take action This is the final step in the process of situation sensing. It requires a tolerance for risk and uncertainty. It is one thing to use intuition to draw a conclusion about current conditions in the operating unit, but a far riskier matter to take decisive action based mainly on intuitive reactions. Yet that is exactly what the exceptional performers do.

The willingness of exceptional performers to accept their intuitive actions is enhanced by the fact that situation sensing is different from "shooting from the hip." Although sensing a situation is an intuitive, nonrational, unstructured approach, it involves considerable study, analysis, reflection, and integration. Managers who free themselves to be guided by intuition have actually done a great deal of spadework before letting intuition take over.

For example, a president of a large company virtually eliminated its middle-management corps because situation sensing and the analysis of an outside consulting firm convinced him that important ideas were being inhibited by the middle-management level. The president first encouraged everyone to bring good ideas directly to the top. This brought no results even after the president made himself highly visible and approachable. Finally the president proved that he had faith in his own intuition by radically changing the organization of the company, creating direct lines of communication from lower-level managers and technical people to his office.

Getting Started

Try rehearsing this situation in your mind. You are an anthropologist with a government grant to study the technology, economy, and social and cultural conditions of the operating unit you manage. You arrive at the office one Monday morning prepared to spend the next several months trying to understand what the people in the department do, how they do it, and how they feel about doing it. How would you go about understanding this foreign social situation?

What would you study first? Who would you talk to? What questions would you ask? How would you satisfy yourself that you actually had come to a good understanding of the operating unit?

Now consider the knowledge you actually have of the operating unit. What information would the anthropologist uncover that you do not know already? Chances are good that some information you have never even considered would be uncovered by the anthropologist. Use the methods you just derived from your rehearsal of the situation to gain this information for yourself.

Now decide what situations need action. There may be none, but probably you will find some. Use the same broad investigative technique that you used to scan the situation. Observe, talk, analyze, and study the real world around you until you get a signal or a gut feeling about what ought to be done. Then . . . do it!

MANAGEMENT PRACTICE 14: SHARE THE GLORY

Definition

The essence of sharing the glory is to unequivocally assign the credit for accomplishment to the individuals and teams that really did the work. This is directly contrary to the common situation where the boss gets the credit even if he or she did no productive work on the project. Sharing the glory is the practice of acknowledging and recognizing a person's contribution *in a meaningful way.* It is a manager's fulfillment of the expectations the manager and the employee established.

Sharing the glory should be built into the system. When it is a standard part of the operating process of a department or company, exceptional performance will be encouraged. Giving appropriate recognition and reward for good work is a critical factor in building long-term relationships with employees. It makes the work more interesting and satisfying.

Historically we have addressed this question of rewarding good work by giving titles and increases in salary and benefits. Companies recognize outstanding employees with banquets, special trips, award ceremonies, special gifts, letters of recognition, and acknowledgment in company newsletters.

Then a more humanistic philosophy emerged, largely during the 1960s. This approach stressed warm, personal human contact as the only meaningful reward for performance. Later yet, a far more equal and participative atmosphere has emerged. Less distinction is made between levels of management. Teamwork and group participation efforts are meant to allow people freely to express their feelings. These approaches are designed to improve employees' ability to work together.

Many managers doubt that anything has really changed, however. "Even though we'd like to think that companies have progressed over the last fifty years from virtually being authoritative to being humanistic, companies still continue to try to get people to 'buy into the company,' " said Dr. Richard Mignerey, CM, director of marketing and development for the Center for the Study of Administration, Nova University.

> "[If companies really reward people humanistically,] what makes companies give people three months off with full pay after six years' employment, or give employees seven-course lunches for only twenty-five cents with first-run movies every other Saturday night, or free unlimited worldwide air travel after ten years of service? What ever happened to the idea that companies don't have to

> entice people who like their job? Perhaps all of these special programs are designed to cover up management's lack of rewarding and acknowledging people in a meaningful way so that they will want to stay."

Differentiation

Sharing the glory is valuable in any business context. Successful high-tech managers find this management practice especially valuable in the high-tech environment.

Both the environment and the nature of the work make personal reinforcement more necessary. The high-tech environment requires people to take risks. They are putting their skills, judgment, and courage on the line nearly every day. When they take a risk that produces good results, they need to hear about it. They need a distinct reward. In the absence of this reinforcement, employees' risk taking will diminish and eventually disappear. Yet risk taking is essential to doing a good job.

Much work in high-tech companies and departments is detailed, complex, impersonal, and tremendously subject to error. The work tends to be socially alienating, even when job tasks have considerable intrinsic interest. The impersonal nature of the work should be compensated for by an unusual amount of personal contact in the management and administrative areas. A good high-tech manager makes up for the social distance of work tasks by reducing the social distance of management contacts.

Importance

It's easy to find a number of factors that make sharing the glory important:

- △ Companies lose valuable managers and technicians when management does not discover and use a meaningful way of sharing the glory. According to the managers in the survey, a frequent reason for leaving a company is employees' feeling that their personal contribution is unrecognized and unappreciated.
- △ When people are given recognition and credit for their ideas and task accomplishments, a sense of loyalty toward the company grows. This makes employees more willing to contribute their best ideas to the common good of the organization.
- △ Sharing the glory is one payoff of subordinates' expectations. Every time a manager and subordinates fulfill their agreements, the bond between them grows stronger. This pride and commitment increase the chances of attaining exceptional performance.
- △ Sharing the glory promotes a creative atmosphere. When individuals are given direct credit for the ideas they share, they relinquish the feeling that they should keep their best ideas to themselves. Employees who fear intellectual pilferage of their ideas with no reward forthcoming will eventually become secretive about their best potential contributions.

How to Do It

Brian Burns points to one answer:

"As my tech people do their work more efficiently, I have found that it is meaningful to

> the employees if I reward their efforts by giving them jobs where they interact with people on the outside. I might have them deal with the person who sent the data and discuss with them alternative suggestions of ways our company could be of help. It could involve handling complaints or helping to find solutions to complex questions."

Not all employees would find it rewarding to deal more with outside clients. The underlying features of this method, however, have a good chance of working with any group of employees in any management situation. The rewards are autonomy, expanded control of resources, and a richer and more varied work assignment. These changes satisfy the entrepreneurial desires of many people inside and outside high-tech positions.

Whenever possible, associate the names of the people responsible with the presentation of the product. Even today you often hear people recall how two amateur radio operators who worked on the Apollo communications hardware went to great lengths to sneak their call signs onto a printed circuit board. They wanted to go into space with the astronauts.

Designers of video entertainment programs and games like their names associated with their products. One designer interviewed said, "The name is just as important as the royalties."

When technicians or managers contribute important work, encouraging more contact with higher management will often be rewarding. Such contact may improve self-esteem and strengthen the person's self-image as an important part of the company. An event as simple to arrange as dinner with the general manager or division head can be quite effective.

A personal memo from someone in higher management can be a meaningful reinforcement. The person writing the memo, however, must demonstrate that he or she is genuinely familar with the contribution the technician or manager has made.

Getting Started

Getting started sharing the glory can be difficult. It is unlikely that you will be able to improve much with a formal program. The real need is to adopt the attitude that whoever actually does the work or produces the idea gets the credit.

One analytical step may help you get started. Review the last six months of operations. Your calendar will probably remind you of the major events and accomplishments of the period. List the successes your operating unit has had. For each item on the list, analyze which people were most directly responsible for the success and what meaningful recognition they got for their contribution.

Chances are that you will find occasions where the people most responsible for a success got nothing in the way of recognition. In some cases they may even have received negative feedback if many problems had to be overcome in a complex situation.

You may also wish to acknowledge the good work that went into failures. People who are really cooking, trying hard to get the most from personal and company resources, are bound to make some mistakes. They may legitimately expect to receive recognition of their efforts even if all goals could not be reached.

MANAGEMENT PRACTICE 15: TARGET PERFORMANCE

Definition

Targeting performance is the process of pointing out to an employee areas of needed improvement. The preferred method is indirect, using collaborative and self-discovery approaches.

Defined this way, targeting performance is obviously an element of traditional process management. It consists of discovering areas in which actual performance does not match planned and desired performance. Then corrective actions are applied to bring performance into line.

Obviously, sharing the glory is a part of this process. Recognition, praise, and credit for good work tend to reinforce and increase the frequency of the desired performance. There is another side of the coin, however. Criticism of ineffective or wrong-minded performance is also needed.

Differentiation

Throughout, we have stressed the desirability of allowing and encouraging an entrepreneurial approach to work, but people who have successfully adopted this point of view will often find it hard to accept criticism. One of the foundations of exceptional performance is a great faith in one's own competence and skill. Direct criticism may be counterproductive in the long run.

"This is something that I have experienced," said Joyce Campbell, former personnel director for one of the country's largest nuclear power construction companies. She said during our interview session, "In a high-tech environment a manager is frequently working with people who are extremely sensitive, and being direct and constructive can result, at the bottom line, . . . in being destructive."

Paula Carter, personnel director for the western region of AT&T, during an interview expressed similar thoughts about the need to approach people in a high-tech environment differently. She said that successful people who view themselves as competent have a hard time accepting criticism.

In a traditional setting, the use of authority will be more acceptable to subordinates because they will be largely used to this approach to begin with. People who are used to taking direct orders will usually be more attuned to accepting direct criticism.

Importance

In any environment, guiding performance through criticizing ineffective actions is one of the most stressful aspects of the manager's job. Managers emphasize that providing this guidance is a stress producer partly because the managers lack expertise in the practice. Stress and lack of perceived skill thus work together to make targeting performance something that many managers avoid.

This is truly unfortunate. When a manager is ineffective at criticizing others, he or she puts the success of the project in jeopardy. Ineffective criticism also threatens the whole atmosphere that the exceptional manager has worked hard to build. It attacks the confidence and sense of mastery that employees and managers feel toward their work.

When a manager avoids criticizing others, employees develop an inaccurate perception of themselves and their work performance. Their images of themselves become inflated and out of proportion. Employees may walk around thinking themselves winners when in fact they are, or are about to be, losers. When performance deficiencies are finally revealed, the outcome is painful and often counterproductive.

Consider, for example, one high-tech company that—because of its rapid growth—kept promoting a certain technical person. For three years his superior never sat down with him to discuss problem areas in performance. Yet as this person's responsibilities grew, performance failures became more and more of a problem.

The company reorganized, and our ineffective technical manager began reporting to another superior. The new superior quickly recognized the extent of the problems and immediately approached the manager about some of the performance difficulties.

The subordinate's first reaction was shock and denial. He genuinely hadn't known that the problems existed. It took months before this manager was able clearly to see and acknowledge his weaknesses. Rather than choosing to work on developing his managerial skills, he decided to return to work in his technical specialty.

What happened to this manager is not unusual. It is especially common in high-tech enterprise where there is rapid growth and fast advancement to higher levels in the company.

How to Do It

In the first place, be sure not to mount a campaign of constructive criticism. Such an undertaking would threaten every success your operating unit is capable of.

The exceptional performing managers in the survey recommend using an "oblique approach" to improve the quality of performance. In this practice, a screen or frame of reference is used for pointing out deficiencies. The emphasis is on directing the criticism to the occurrence or behavior rather than toward the person. This is often called criticism *ad rem* rather than *ad hominem*. The successful managers revealed several ways to do this.

1. Compare performance or results with a concrete standard For example, assume that an important industrial customer has just returned an entire shipment of tools. One manager might report this to the subordinate in charge this way: "U.S. Manufacturing returned our entire last shipment. You must have missed something. Go back and find out what's wrong and tell me what we need to do." Another manager might say, "U.S. Manufacturing returned our last shipment. They said the tools didn't meet the specs on the drawings with the purchase order. Look into the situation and report back by four o'clock to update me about what's going on."

There is only a subtle difference between these two managers, but it is an important one. The second approach is more oblique; at no point did the manager directly criticize the subordinate for ineffective behavior. The manager only pointed out that the shipment did not meet purchase-order standards. The subordinate is free to buy in to the situation as a responsible and competent problem solver.

Other indirect phrases a manager can use are "not up to par," "not up to standard," "not up to specifications," and "not coming through as ordered."

2. Mirror the problem Charlie, who manages data processing for a medium-high-technology company, uses this approach effectively. If an employee has made errors on a program, Charlie will run the program himself. He will then go back to the programmer and say, "I've been using this program and have been running into some difficulties. How about you? Have you found these same problems?" Charlie presents the programmer's problem as if it were his own. Then by collaboration the difficulties are resolved.

Another mirroring approach involves recreating the situation using someone else. If Charlie were to use this approach, he could have said, "I was talking with Tom at XYZ Company yesterday and he was saying that when they used the same version of your program we use they ran into some difficulties with the sort routines. Is this something you've encountered?"

3. Use the referral approach Like the other indirect methods, referral emphasizes a collaborative effort. Referral is especially valuable when a subordinate needs criticism or help with a problem but the manager has insufficient technical expertise in the area to feel comfortable giving guidance. Referral is also helpful in a creative endeavor when a subordinate loses focus on what's most important.

In referral, the manager teams the subordinate with another employee who has expertise in the area of concern. The two then work as a team to solve problems and resolve performance deficits.

During our interview, President Harold Shoup described a practice used successfully at Carr Liggett:

> "One of the most effective ways to target performance for the creative person is to have someone in the loop whom everyone respects. Managing that creative person is a challenge, and there are times when a creative person gets hooked on an idea. By bringing in someone who has the respect of that person, everything quickly gets resolved. When the respected person tells the creative worker if he or she is off target it is easily accepted. In our business, respect is the key word. It's more

important than finesse."

The situation in which a manager is faced with employees with superior technical skill is common. It can also present a serious difficulty. One high-level manager at an automotive company revealed confidentially that because he lacks technical expertise, he directs conversations with subordinates into areas that he is comfortable with. As an example, if one of his managers comes in with a technical problem and he begins to feel uncomfortable, he takes the offensive by starting to talk about marketing. He is an expert at marketing and knows that his subordinate managers are not. The managers admits that this response is ineffective, but he has not found a better way to deal with the situation.

The referral approach is an effective alternative. A good practice is to express the referral as a need for help: "I need your help, John. Here's the standard we need to meet. You and Clarissa have the expertise. Let's get this worked out." Now the manager's main job is to keep efforts focused on the critical performance factors he or she is interested in.

Getting Started

You should cover the matter of giving criticism when you first discuss expectations with subordinates. Ask them directly how they best receive criticism: "There are going to be times when you make mistakes or when your performance will not be exactly what we need. How would you like me to express criticisms I have?" Most employees will be able to give you useful information in response to this question.

The best policy for using an oblique approach to criticism during day-to-day work is simply to prepare yourself before you take any action to target performance. In trivial matters, prior preparation is usually not necessary. But in any matter of importance, decide in advance how you are going to express criticism. Choose the method of comparing to standards, mirroring, or referral and have an idea of the exact terminology you will use.

MANAGEMENT PRACTICE 16: MAINTAIN SANG-FROID

Definition

Sang-froid means the display of control and grace under pressure. Literally the word means "cold blood." Its implication is that emotional factors—anger, fear, discouragement—should not interfere with carrying out work and dealing with others in the company.

Maintaining sang-froid might be said to have two aspects:

1. It is important to *appear* calm and cool during crises even if you do not feel that way. Whatever is your own equivalent of counting to ten before you speak will be useful in developing this capability.
2. Over a period of time, you can develop the habit of actually *being* calm under fire. This ability is closely related to years of experience on the job and having built the kind of strong, dedicated staff that give you confidence in their abilities.

Sang-froid definitely does not mean an indifference to the concerns and fears of others. On the contrary, if you can maintain your wits and objectivity, you will be able to deal more effectively with others.

Differentiation

The loss of sang-froid will nearly always cause a deterioration in quality and quantity of work. This deficit has a greater tendency to produce failure in the high-tech environment than in more traditional, procedure-oriented firms for two obvious reasons:

- In high-tech, there is a greater need to be first. If the pace of a project is interfered with by stray emotions, the company may well lose the competitive advantage of hitting the market first. This can have serious consequences.
- The loss of sang-froid often causes a serious loss of emphasis on quality. When everyone is running around saying, "What are we going to do now?" the proper emphasis on control of the product is lost. Stress is directed toward doing *something* rather than on doing *the right thing.*

Importance

The manager is a leader. One characteristic of this role is its symbolic value. People take readings of the emotional state of the leader and convert their observations into a generalization about current conditions. If the leader is seen to be upset or overly discouraged, subordinates will believe that the situation is grave. Conversely, in even very rough situations, subordinates will take heart if the leader is seen to be rational and confident.

When the leader appears to be upset, subordinates' performance will suffer. An important reason for this deterioration of output is the simple misdirection of energy. If everyone is trying to work out just how bad things are and speculating about how the problem will affect him- or herself, the time and energy for doing good work are lessened.

Further, the fundamental process of exceptional performance—confidence, pride, and commitment—will suffer seriously. Employees will wonder whether the problems are their own fault. The organization with which they identify will be seen to have serious, threatening problems. Pride in work well done will lessen because employees will begin to question whether the work *has* been well done.

How to Do It

First, recognize that sang-froid is a skill. It can be learned in the same way you learn to ride a bicycle or sink a golf putt. To stay cool when faced with difficulty, it is necessary to integrate two levels of response to a bad situation.

1. Consider the broad perspective Look at the overall organization; is your current problem really likely to sink the whole ship? Considering the whole situation of which your problem is a part enables you to determine if the problem is as serious as it first appears.

Some people keep situations in perspective by believing—as held by Eastern philosophies—that we enter this world with nothing and leave it with nothing. Anything in between is mainly our choice. Others practice the technique of looking in the mirror and asking, "When you are seventy-six years old, what do you want to remember and be proud of in the life that has passed?" Yet others remind themselves, "Nobody is perfect!" These people genuinely accept the literal truth of that phrase. If things aren't going smoothly, if they have made an error in judgment, that's not necessarily horrible and disastrous. Everyone they are dealing with has made similar or worse errors.

Being able to keep situations in perspective helps produce exceptional performance. Overreacting to a problem or error paralyzes our ability to deal with the current situation. The exceptional managers in the survey recognize that they have the element of choice in every situation. As a result, they are led to believe more strongly in their decisions and in the actions they take. This strong belief is felt and responded to by others.

2. Keep in mind what is most important on the specific action level Integrate the broad perspective with the current arena in which you take action and make decisions. The exceptional performers recommend consciously acknowledging that to a great extent managers are in control of all work situations. Recognize that periods of chaos and uncertainty are inevitable in any enterprise that depends on creativity and innovation. Accept these periods as transitional; don't push people for absolute certainty. Don't make a big deal over issues that really have low priority. Plan and take steps that relate to the truly important factors in the situation.

Integrate life experiences and outside learning to help stay composed during highly stressful periods. This will aid in keeping an optimal level of performance.

During grueling meetings or interviews, do whatever is possible to make yourself physically comfortable. Learn to use a breath control or mental imagery technique to calm yourself.

Exceptional managers do not see heavy work loads as a personal threat; they recognize that pressure is generated internally. Pressure is usually the manager's fear of not having enough time or resources available to meet his or her own expectations and the expectations of others. A project or task has no inherent pressure: it never did and never will. The interpretation of that project or task is what creates the pressure. Knowing that pressure is self-induced enables managers to take responsibility for controlling it. They are aided in dealing with pressure by separating the feelings of pressure from the task or project itself.

Simple responses to pressure often work wonders. The exceptional managers may take short breaks, engage in physical exercise, or use various relaxation techniques. They base their task-related actions on consideration of control, delegation, levels of expectation, and task importance.

Exceptional managers have recognized that consciously working to maintain calmness is important to helping the operating unit achieve end results. To do this, the manager must recognize and deal with his or her own emotions and the emotions of subordinates, peers, and superiors.

Getting Started

Developing sang-froid is a long-term undertaking. These suggestions may speed the process:

- Analyze the pressure you are under right now. What is the true source of the pressure? It must be coming from your own expectations or from the expectations of others. Mentally separate this pressure from the objective difficulties of getting projects or tasks done. To accomplish a task will take *x* amount of work, no matter whether you are pressured or not. Deal objectively with the pressures. Ask for more time or resources. If the pressures cannot be removed in that way, find ways to deal with the emotions that the stress is causing.
- Learn how to relax. Specific techniques for relaxation, isolation of stress, and self-analysis can be very helpful. Books, articles, and workshops are available for learning these techniques.
- Forget the idea that you *need* to feel pressure. The main steps to be taken remain the same whether you are happy or unhappy. Feeling strong stress is not necessary for optimal performance.

ACHIEVING EXCEPTIONAL MANAGERIAL PERFORMANCE 10

Summing Up

Robert Felts recently had his first airplane ride. Mr. Felts journeyed to East Texas to confer with a customer of Felt's employer, Bethlehem Steel at Sparrows Point, Maryland. The reason Mr. Felts had done little business traveling for the company is that he is not a marketing manager or applications specialist, but a veteran production-line worker.

At about the same time, Westinghouse Electric sent Clay Adams, a machine operator, to Japan to observe conditions there. Mr. Adams was strongly impressed by the competitive spirit of the Japanese and came away with the conviction that "the Japanese aren't going to slack up."

Westinghouse also recently sent thirty hourly workers to New York City to evaluate the performance of their products in the city's transit system. Jack Geikler, manager of Westinghouse's transit divsion, says, "There's a difference between putting wires into a black box and riding the product through the South Bronx."

The Bethlehem Steel program has produced a few jolts for the employees who have visited major customers. One foreman at a customer's plant told a Bethlehem worker, "You guys at Sparrows Point are garbage rollers." The employee said that statement really hurt his pride.

Other Bethlehem workers recall the shock they felt when flaws in the steel they had shipped showed up in buyers' products. One major customer was ready to shift to a different supplier because of continuing flaws in Sparrows Point plate steel. Bethlehem management didn't seem to be able to do anything about the problem. Finally, a group of Sparrows Point plate workers was sent to the customer. When the steelworkers saw the problem their defects were causing, they were "practically in tears," the buyer said. The problem was solved overnight.

What these examples have in common is that they give employees a chance to be truly involved with the success or failure of their company. The workers do not just attach wires and put in nuts and bolts; they are given the chance to see the true importance of their work. They have the chance to be engaged in the world where their organization competes. They are encouraged to provide dedication to quality as a result of their personal pride and motivation.

This pride and personal engagement are exactly the results sought by exceptional managers. The exceptional performers in the survey strongly believe that when people are given

- an understanding of the competitive situation,

- a genuine knowledge of the company's vision and major strategies,
- the skills, knowledge, and confidence needed for doing their work,
- the resources for outstanding performance,
- sufficient autonomy to do the work in the way that is best for them.

the employees' work will be outstanding. Creating this climate for exceptional performance can be accomplished through conscious and intuitive intervention in the organization.

The question this final chapter answers is "What now? I've read this book. I understand the sixteen management practices. What do I do next?"

As with any other management action, you need a plan. Certain actions will help you implement the important practices on the job. The key word is "action." At this point, you want to *do things* to let exceptional performance take a firm grip in your operating unit.

The following pages feature a planning form that will help you get started. The form invites you to be definite about how you will phase in the recommended management practices in your operating unit. The form also includes a schedule for implementation of suggested activities, so that you can begin applying them in an orderly way.

The form lists some actions that you can use to begin applying the management practices. Following this, you are asked for five specific pieces of information:

1. the planned date upon which the action will begin
2. the date the action was actually completed
3. expected outcome

4. a personal rating of the success of the action
5. comments describing the effect of the action in your operating unit

These heads are self-explanatory except for the success rating. Use some scale for indicating what you perceive as the relative usefulness of the action taken. Some managers use letter grades A to F, as in school. Others use a scale from 1 to 10, as in scoring some sports performances.

I recommend that at first you select only one management practice to develop. Commit yourself over the next two weeks to constantly work on the chosen practice so that it is internalized as part of your personal management style. This beginning will direct you toward doing things *better*, as opposed to simply doing *more* things.

IMPLEMENTING EXCEPTIONAL PERFORMANCE PRACTICES

Management Actions	Date Begun	Date Completed	Expected Outcome	Success Rating	Comments
1. Provide experiences to learn from. • Plan to include goals, training, valuable experiences through meaningful activities. Establish a schedule for each in the next appraisal interview. • Remember to stress the learning value of each formal or informal work assignment.					

Management Actions	Date Begun	Date Completed	Expected Outcome	Success Rating	Comments
2. Practice letting go.					
• Use appraisal system information combined with personal observation to list the strengths and weaknesses of each person who reports to you.					
• Design a single delegation for each employee as an experiment. Remember goals and standards and examples of good and bad work.					
• Analyze carefully the resources needed to accomplish the delegated tasks.					
• Take actions necessary to provide direct control of the resources to subordinates as needed.					

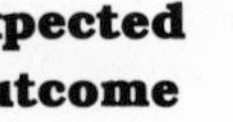

Management Actions	**Date**		**Expected**	**Success**	**Comments**
	Begun	**Completed**	**Outcome**	**Rating**	
3. Use symbols and slogans.					
• Review the current orientation program with an eye toward improving its communication of the rituals, norms, mores, symbols, and informal history of the organization.					
• Check at least weekly to see that you have made a serious effort to let all employees know what is going on in the company.					
• Make plans for at least one group ritual in your operating area. Attempt to match the interests of employees.					

Management Actions	Date Begun	Date Completed	Expected Outcome	Success Rating	Comments
4. Create the excitement to achieve. • Use introspection to identify areas where your own feelings may stand in the way of exceptional performance. Try to plan specific actions, like a meeting with your boss, that will increase your own excitement to achieve. • Consider the recent accomplishments of the people who report to you. Can you find any outstanding work for which the person producing the work got inadequate praise and/or credit? • Make a schedule to provide the necessary, genuine praise that builds commitment.					

Management Actions	Date Begun	Date Completed	Expected Outcome	Success Rating	Comments
5. Identify the expectations operating within the organization. • Choose a single situation in your organization and identify the expectations involved in it—both spoken and unspoken. Determine the source(s) of those expectations. • Before sharing the identified expectations with others, describe in your own words (a) the true vision pursued by the organization and (b) the methods of working together that are genuinely supported by top management.					

Management Actions	Date Begun	Date Completed	Expected Outcome	Success Rating	Comments
6. Make expectations explicit. • Choose a single important task to accomplish in your unit. • Arrange and hold a meeting with the person to whom you will delegate the task. • Attempt to convey your expectations and those of the organization. • Try to learn the expectations of the person you wish to commit to accomplishing the task.					

MANAGEMENT ACTIONS	Date Begun	Date Completed	Expected Outcome	Success Rating	Comments
7. Build the expectations package. • Try again, having had the experience of negotiating expectations under management practice 6, to express in writing very explicitly your own true and complete expectations about the people who work for you. Consider every person who reports directly to you. Make the list as exhaustive and accurate as possible. • Devise specific events and ways of working that will satisfy the important expectations of everyone involved.					

Management Actions	Date Begun	Date Completed	Expected Outcome	Success Rating	Comments
8. Communicate a broad vision. • Review written plans to determine if they communicate a strategic vision of what the company is trying to achieve. If plans are entirely numerical and narrowly focused on quantitative results, revise them so they also communicate a view worthy of personal dedication. • Talk with subordinates, peers, and superiors purely to discover ideas worthy of inclusion in a vision of the corporate direction. • Review in your mind whether you truly understand and support the major strategic aims of your organization. • If some areas are unclear, meet with your boss to clarify them.					

Management Actions	Date Begun	Date Completed	Expected Outcome	Success Rating	Comments
• Work out in writing simplified, capsulized versions of the major elements of the vision using slogans and symbols.					
• Use these when communicating with others in your operating unit.					

Management Actions	Date Begun	Date Completed	Expected Outcome	Success Rating	Comments
9. Focus sharply on what is important.					
• Use complete strategic plans plus personal analysis to decide which tasks and accomplishments are truly the most important. Devise clearly defined events related to these accomplishments.					
• Identify the important people who have the authority or influence to affect the accomplishment of the most important goals. Plan events to gain their support.					
• Identify the key subordinates and peers who will contribute the most to the accomplishment of the important goals.					

Management Actions	Date Begun	Date Completed	Expected Outcome	Success Rating	Comments
• Plan events that will allow these subordinates and peers to approach the work in a way that is meaningful to them.					
• Continue to communicate the symbols and slogans that have been established to inspire subordinates and other important people to stay focused on what is most important.					

MANAGEMENT ACTIONS	Date Begun	Expected Completed	Success Outcome	Rating	Comments
10. Act decisively to empower others. • Consider all the people who report to you and judge their readiness to accept full authority for pertinent tasks. Provide events that will increase autonomy, confidence, and skill where needed. • Choose a task that is important but not critical. Plan a first event to give a subordinate the chance to buy in to the project. Plan a further series of events to provide resources and more detailed information about the project and clarify expectations. Give the employee full responsibility for completion of the task.					

MANAGEMENT ACTIONS	Date Begun	Date Completed	Expected Outcome	Success Rating	Comments
• Give full credit for results to the people who actually did the work, took the initiative, and accepted the final responsibility.					

Management Actions	Date Begun	Date Completed	Expected Outcome	Success Rating	Comments
11. Communicate a sense of competitive urgency. • Improve your own knowledge of the competitive situation of the organization. • Analyze the relationship of your own operating unit to the overall competitive success of the organization. • Continually communicate the competitive importance of tasks and projects. • Plan events to keep internal competition at a desirable level.					

MANAGEMENT ACTIONS	Date Begun	Date Completed	Expected Outcome	Success Rating	Comments
12. Create activity and momentum (make it happen). • Review your plans and actions up to this point. Check again that you know what's truly important. Are your planned events successful because they are "making things happen" consistent with the direction of the group/organization? Check that the important people you chose were the right ones. • Assess the likely risks of taking the initiative yourself. • If the risks are acceptable, start pushing.					

MANAGEMENT ACTIONS	Date Begun	Date Completed	Expected Outcome	Success Rating	Comments
13. Sense the situation. • Circulate among the work areas of the operating unit you manage. Use your intuition to interpret what you see. • Consider which areas, tasks, and projects call for your action. • Talk to others, analyze observations, and study the real world of operations to get a strong intuitive feeling for what needs to be done. • Then do it.					

MANAGEMENT ACTIONS	Date Begun	Date Completed	Expected Outcome	Success Rating	Comments
14. Share the glory. • Review the last six months of operations to find instances of success. • Decide which people were directly responsible for success and what meaningful recognition they got for their contribution. • Use the improperly credited situations you discover as a guide for sharing the glory in the future.					

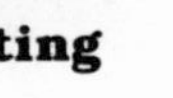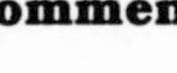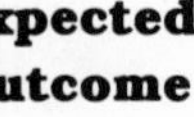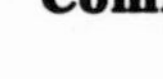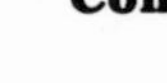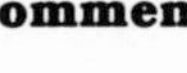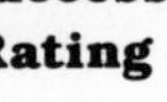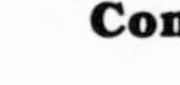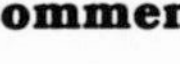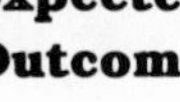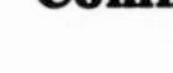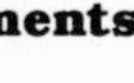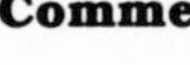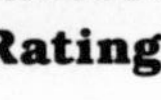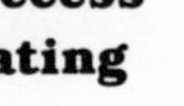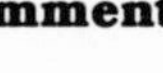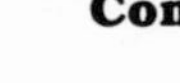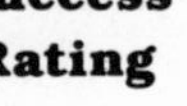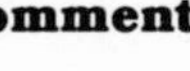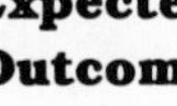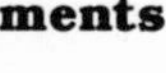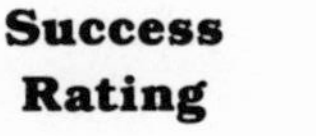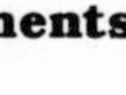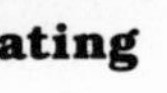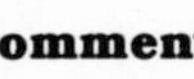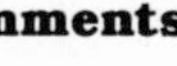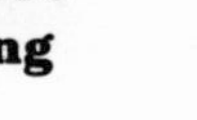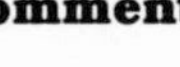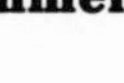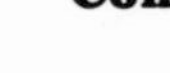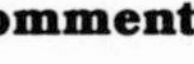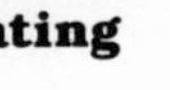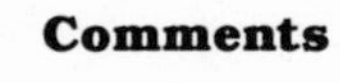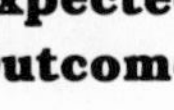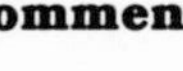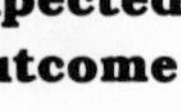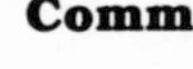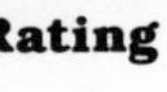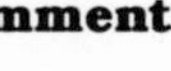

Management Actions	Date Begun	Date Completed	Expected Outcome	Success Rating	Comments
15. Target performance. • Review your orientation program; plan future discussion(s) of how each new subordinate prefers to receive criticism. • In important matters, prepare yourself to give criticism that is consistent with the expectations established. • Choose the method best suited to the individual(s): comparing to standards, mirroring, or referral.					

Management Actions	Date Begun	Date Completed	Expected Outcome	Success Rating	Comments
16. Maintain sang-froid. • Analyze the true source of pressures you are under now. Deal objectively with the pressures. • Learn one of the numerous effective ways to relax. • Judge whether you feel the *need* to be pressured. If so, analyze whether you really do better work under pressure or whether you get other satisfactions from pressure and the accompanying high stress.					

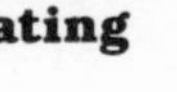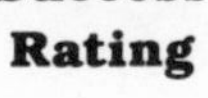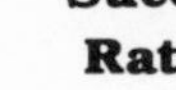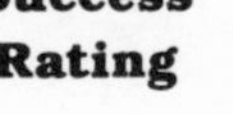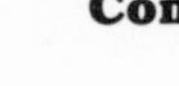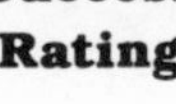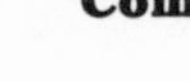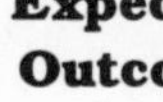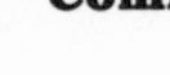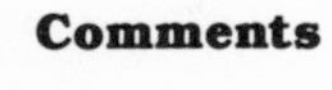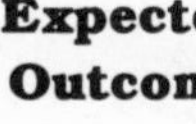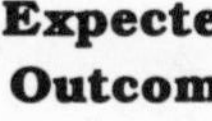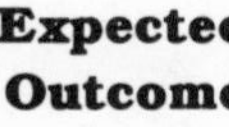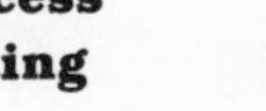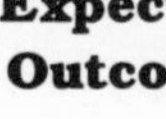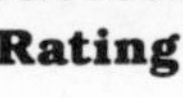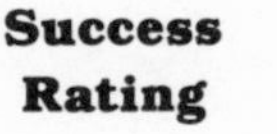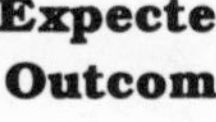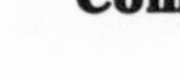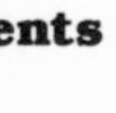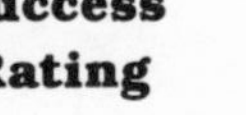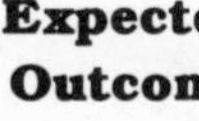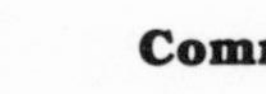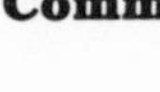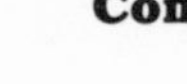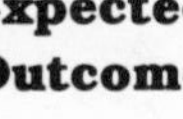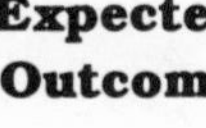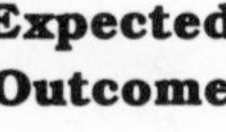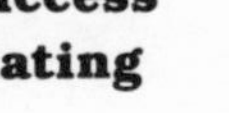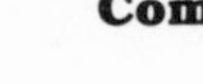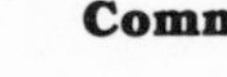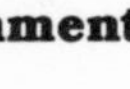

APPENDIX: NOTES ON THE METHODOLOGY OF THE NATIONAL HIGH-TECH MANAGEMENT SURVEY

My first effort consisted of a traditional management values and practices questionnaire administered to over three thousand managers in high-tech and traditional industries.

Briefly, this first National High-Tech Management Survey obtained responses from managers at various levels of responsibility at major firms such as AT&T; Ford Motor Company; Gould, Inc, Computer Systems Division; Burger King; the American Farm Bureau Federation; and Pacific Gas and Light. Further, it incorporated a national sampling of managers who are members of affiliates of the National Management Association. The executive committee of the Massachusetts High Technology Council was also surveyed to obtain more detail on the special viewpoint of chief executive officers.

The sample was designed to yield two major comparisons:

- High-tech managers as a group versus managers in traditional industries.
- Exceptional performing high-tech managers versus other high-tech managers not identified as exceptional performers.

The exceptional performing high-tech managers were chosen for inclusion in that group by upper-level managers in the high-tech companies and departments.

The population of high-tech managers, collectively, that were identified in the study was not found to differ significantly from managers in traditional industries on the attitudes and beliefs about management measured by the questionnaire. The results showed, indeed, that the entire sample of managers had been exceptionally well indoctrinated in the current academic emphasis on "people management" as an essential extension of the traditional management functions of planning, organizing, staffing, and controlling.

Comparisons between the exceptionally performing high-tech managers and the other group of high-tech managers yielded nearly the same results on the factors measured by the questionnaire. A slight difference in emphasis did emerge between the exceptional performers and the other high-tech managers in some areas. The exceptional performers, for example, stressed "developing an effective communications strategy" and "meeting corporate objectives and goals" measurably more than did the managers not chosen by their superiors and peers for inclusion in the exceptional group. Other responses appeared to indicate that the exceptional managers were better informed about the organization and about what the leadership wanted.

My response to these preliminary findings was disappointment. In my personal discussions with hundreds of managers, nearly everyone had agreed that legitimate differences were likely to be found between managers in high-tech industries and those in traditional industries. Yet over three thousand questionnaires had clearly failed to show that significant difference.

Either the widespread belief in a difference was mistaken or the questionnaire had failed to focus on and elicit the kinds of beliefs and behaviors that reflected the difference. If the latter possibility could be eliminated, then the first conclusion—that there was no real difference—would have to be accepted.

To further resolve the issue, I took a fresh investigative tack by adopting a new method of gathering information. A printed questionnaire, even the best-written one, limits the range of responses that survey participants can give. However, this limitation, when the field of study is already well defined, is usually more than compensated for by the quantitative strength of the results.

Another study approach, commonly referred to as the clinical method, avoids the limits imposed by a written questionnaire. In clinical data collection, the investigator uses face-to-face interviewing. Open-ended questions can elicit responses from the entire spectrum of opinions, attitudes, knowledge, and practices of members of the survey group. At the same time, preplanned questions and areas for follow-up can impose at least a measure of structure on the interviews and provide a firmer base from which to make later generalizations. These are the advantages often gained by sacrificing the numerical rigor available with a written survey tool.

I chose the clinical method for continuing the investigation of management beliefs and practices of high-tech managers. Thus it came to be that the collection and analysis of thousands of written survey responses were followed by the administration and analysis of hundreds of face-to-face interviews with members of the same universe of managers.

NAME INDEX

COMPANY INDEX

ABOUT THE AUTHOR

As a management consultant, Deborah Bright helps a wide variety of business organizations bring about positive internal change. She works with key managers in improving their overall job performance and conducts workshops in the fast-lane management practices she describes in this book.

As a sports consultant, Dr. Bright works with professional athletes in maximizing their performance in highly competitive situations. Once ranked among the top ten U.S. women divers, her impressive career in platform and springboard diving led to competition in the Olympic trials.

Dr. Bright is an adjunct professor at Wayne State University and Nova University, where she teaches graduate and undergraduate courses in stress and human resources management. She is the author of *Creative Relaxation: Turning Your Stress into Positive Energy* and the three-hour audio seminar program, *Gearing Up for the Fast Lane: New Tools for Management in a High-Tech World.*

A NOTE ON THE TYPE

The text of this book was composed, via computer-driven cathode ray tube, in a type face called Bookman Light. The original cutting of Bookman was made in the 1850's by Messrs. Miller and Richard of Edinburgh. Essentially the face was a weighted version of the popular Miller and Richard old-style roman, and it was probably at first intended to serve for display headings only. Because of its exceptional legibility, however, it quickly won wide acceptance for use as a text type.

This book was composed by Rogers Typesetting Company, Inc., Indianapolis, Indiana.
Printed and bound by Halliday Lithographers, West Hanover, Massachusetts.
Text Design by Positive Identification, Inc., Indianapolis, Indiana.
Cover design by M.R.P. Design.
Production supervised by White River Press, Inc., Indianapolis, Indiana.